CHESS FOR BEGINNERS

A COMPLETE GUIDE TO LEAD YOU TO VICTORY! CHESS FUNDAMENTALS, RULES, STRATEGIES, AND SECRETS FOR THE SUCCESS OF EVERY GAME

Fabiano Carlsen

Table of Contents

Introduction

If you're reading this guide, you've probably been drawn to the game just like the rest of us. That's understandable; chess is a pastime that can be as deep or shallow as you want. You could end up devoting as little time with a game or two a week or become an avid player engaging in multiple games per day. A few end up committing most of their lives to the game, but whatever you put into chess, you'll always get a return on your investment.

What is it that makes chess so appealing to so many people? The answer may vary depending on who you ask. Some find it to be a calming hobby, a chance to shut out the noise of the outside world and lose oneself in the infinite depth of the game. For others, it's a breakneck competitive outlet and an opportunity to feel the rush of personal improvement. However, no matter who you are or why you play, understanding the fundamentals of the game is a sure way to enhance your experience. Just as in any kind of art, sport, or craft, a trained eye is capable of interpreting and enjoying more details in every aspect of the game.

That's where "Chess for Beginners" comes in. With this book in hand, you will be gaining a valuable advantage in understanding the many complexities of chess. As you read from the game's history to its evolving modern strategies, you'll come to fully understand why this game of 32 pieces and 64 squares continues to fascinate people and cultures all over the world.

History of the Game

The precise origins of chess remain a mystery, with historians and anthropologists still debating over the subject. However, what's generally accepted is that the oldest known ancestor of the game originated in India sometime before the 6th century CE. This early predecessor, called "chaturanga," was quite different from the game we know today. A war game, "chaturanga" took its name from a military formation mentioned in the epic "Mahabharata." The formation itself refers to four divisions within the army: infantry, cavalry, "chariotry," and "elephantry."

As "chaturanga" evolved, so did the names by which it was known. Around 600 CE, "chatrang" became a growing pastime in Persia and Central Asia, where it later spread to further east, gaining recognition with different cultures, calling it different names. In Mongolia, it was called "shatar," in China, "xiangqi," and in Japan, "shogi." Each culture brought its unique perspective to the rules of the game and the character of the pieces, but two fundamental qualities of "chaturanga" persisted in each variation. First, unlike in checkers, different pieces had different capabilities. Second, capturing the opponent's king was the path to victory. These qualities remain fundamental to the DNA of modern chess.

Chess reached Europe by the 10th century CE through expanding the Islamic Empire. When "chatrang" was introduced to the Arab world, it was redubbed "shatranj," but remained very similar to the Persian variation of the game.

The early Islamic conquests brought bloodshed to both the Levant and the Iberian Peninsula but also brought cultural and technological innovations, including "shatranj." The Greeks called it "zatrikion," while in Spain, it became known as "ajedrez." Both cultures initially retained the Persian names given to each piece. As the game spread throughout the medieval world, the Persian word "shah" ("king") gradually evolved into English chess. The phrase "Shāh Māt!" ("The king is helpless") would likewise develop into the modern term "checkmate."

Chess quickly took the European world by storm as it became so popular that at times both the church and secular authorities attempted to prohibit the games—and the gambling that often came with it. Eventually, the names of the pieces began to change to reflect the local culture. Elephants became bishops and the "vazīr," or minister, became the queen.

The rules of the game continued to change as well. By 1300, an addition had been made to the rules, allowing pawns to move two squares on their first move. Later, sometime before 1500, the previously weak queen and bishops acquired new abilities to make them more useful and speed up the game. This modified set of rules, once referred to as Queen's Chess, became the modern standard of play by the 19th century.

Since the birth of modern competitive chess in 1851, when German Adolf Anderssen won the first international chess tournament, the sport has become a worldwide phenomenon. Countless grandmasters, men, and women hailing from dozens of countries around the world have risen to prominence throughout the decades.

In the late 20th Century, chess even became a topic of a heated political conversation. When the 1972 World Chess Championship pitted American prodigy Bobby Fischer against Russian champion Boris Spassky, both of the rival nations took immense interest in the match's outcome. When Fischer won the match, ending 24 years of Soviet dominance in competitive play, it was touted as a coup against the USSR itself. Later, when Fischer defied U.S. sanctions to attend an unofficial rematch against Spassky in 1992, a warrant was issued for his arrest.

Today, the Fédération Internationale des Échecs (FIDE) acts as the governing body of competitive chess worldwide. New names have come to prominence in recent years, including Hungary's Judit Polgar, widely considered as the strongest female player in the history of the game. The current reigning champion, Magnus Carlsen of Norway, holds the highest peak classical rating in history with a score of 2882.

Using This Guide

The purpose of this book is to introduce new players to the game: its mechanics, strategies, and the joys of playing. It's meant for both brand-new players and those with some knowledge who are looking to deepen their understanding.

We recommend that you read each section sequentially from start to finish. While all terms are outlined when they first appear, you can return to the glossary at the end of this book for a quick review. Otherwise, the book makes the most sense when each chapter is read in order.

You will familiarize yourself with how to read algebraic chess notation and learn how to read diagrams. You'll also learn the movement options for each piece, as well as a brief overview of the common strategic uses of each piece. Finally, you'll receive an overview of standard piece values as determined by a majority of strategists.

With this knowledge in hand, you will then progress to Chapter 2: The Goal in Chess.

Finally, in Chapter 3: Tactics to Support Your Strategies, chess club, or get started playing online, the final section of this guide provides further study tools as you become a more advanced player. From reading lists to databases, there's an endless, ever-growing selection of chess knowledge out there, just waiting to be studied.

So, let's get started. Today, you will be taking what may well be your very first steps as a chess player. Someday, with dedication and continuous study, you might find yourself writing the next great book on chess! It's your path to pave, and whether you're just looking for a fun new hobby or already have your sights set on grandmaster status, this guide is here to help you.

Chapter 1. Foundations and Rules of Chess

Explanation of Chess Pieces

Pawn

Derived from the word Pawnee who were famous warriors and hunters of the vast wilderness. Such warriors were recognized by the Sioux and were described aptly by the Cheyenne as "The Huntsmen of the Plains."

The pawn is one of the most unusual pieces on the board because of the way it moves and captures pieces. Many people see the pawn as an inconsequential piece, but, when used correctly, it can be one of the strongest pieces you have in your army. Pawns can only move one square at a time except for their very first move when they can go two squares. To capture another piece, the pawn has to do so diagonally, never forwards, backward, or sideways. If the pawn is blocked by another piece in front, it cannot move past it, nor can it take it.

The greatness of the Greyhounds of the Pawnee was first noticed by the Spanish in the 1600s, and they were the first of the Great Plains Indians to ally themselves with the Spaniards at that time. In time, the Greyhounds became the least populous of the Pawnee, but they were also the most feared from the Pawnee war parties. Many small groups of fugitive Indians and merchants were attacked while breaking up the immense expanse of the Great Plains. These attacks took place at the command of the Greyhounds, who treated the Latrine of their rivals, the Kiowa, like a pack of foxes did to the prairie chickens. Men who were living were gagged. Men who were wounded were mercilessly slaughtered. Women and children were taken captive by the Greyhounds. They would skin the captive's animals and leave only the meat. The great strength and prowess of the Greyhounds was nothing more than legend until the events of 1879.

Rooks

Rooks can usually be overlooked by their enemies, who will underestimate their strength and think that their mere image is enough to beat an enemy. The Sioux, however, were capable of observing the slightest detail and would know when an attack will commence. It was very common for enemies to believe that the Sioux would not raise the alarm regarding troop movements or that they would be wary around their village. The Sioux ruse was also recognized, and sometimes when the Sioux were ready, the attack had already begun. The Sioux would call out to their horses earlier, and their songs reached the ears of their horses, hawks, and dogs. They would also keep another eye on flora and fauna, which would make one smell the Americans even before they were spotted.

Knights

Knights are one of the more agile pieces in the game of chess and are, for the most part, on equal terms with other pieces. The best thing about knights is that they can move in the open and jump over other pieces without any constraints. The Sioux were also very capable of doing the same. One incident regarding the Sioux encountered a telegraph crew, and they intercepted the messages and used the telegraph to send false messages back to the telegraph station. They were able to stay undetected for days.

Bishops

Bishops stand by using knights. The tops of these pieces are typically in the form of a hat worn via a bishop (consider that!). Bishops move like rooks, and that is due to the fact they could shift any number of areas, but they circulate diagonally alongside the period of rows/columns, such as rooks circulate. Bishops are given the liberty to move as many places as they need, but only in a diagonal role. For this reason, the bishops have their limitations because they usually relax on a square of the equal shade wherein they started the game. However, their scope is confirmed to be an advantage.

Knights have the most uncommon moving conditions of any chess piece. They are confined to the movement of one or two spaces vertically, one horizontally or one space vertically, and two spaces horizontally, which makes their final movement "L" (that's what it seems like). This precise movement pattern has each a disadvantage and an advantage for both the participant and the opponent, as it permits for a completely unique and meaningful design, which can't continually be avoided; however, this may also be a problem. Just try to create more while you retire.

Bishops are one of the most underrated pieces in the game of chess. The bishops largely keep the enemy king under direct attack. They are also used strategically in positioning the enemy king into the center where the opponent can be more effectively attacked. Bishops will then use the element of surprise and destroy their opponent's king. The Sioux were very capable of this and were famous for their Trickster tradition. It was because of this tradition that they got the nickname "Ghost Dance." The Sioux trickster trickery included wearing various masks and disguising themselves as enemy troops. Another instance was when a Sioux warrior dressed up as an old man and got quarters by the Union troops. He stayed behind enemy lines for many months and asked if there was a Christian among the troops. When one soldier said yes and exposed himself, the Sioux took action.

Kings

Your king is the biggest piece on the chessboard and goes to the bishop on your proper. The king moves in any direction, however handiest one row may be at a given time, making it a distinctly weak piece in terms of movements, nevertheless the strongest element on the chessboard with the least time for the reason that kings seize, and that is how you could win the game.

The most powerful figure in the game of chess. It has no constraints. The most high-strung chess game will be won by the player with the king. The Sioux were such a force that they were able to pinpoint the king of their opponents, and they would cause dissension in the halls of their enemies. They were also very capable of choosing the time and place for the fight. The Sioux would also take the enemy into their lands and "fight behind their lines," hence attacking them where they least expected.

Queen

The Queen is a very complex piece in the game of chess. She is the most powerful chess figure. She can move in all directions indefinitely, and she is able to maneuver as if she has the same freedom as a knight. Once she has found her pathway, she is able to move in any direction until she has reached the square she is to occupy. The Sioux had many queens. They had the mother, the wife, the widow, the daughter, the sister, and the grandmother. They were forced to live their lives in a man's world. They had to withstand the atrocities of the men. They were forced to follow the same rules and regulations and conquer the same territories. They were called upon to defend the lodge, raise the children, make the movable lodge secure, and to tame the horses.

1. **How Did the Monarch/King (If You Are Discussing a Monarchy) of the Greyhounds Deal with Native Tribes?**

Upon assuming his power over the Greyhounds, he planned a way to deal with the other races inhabiting the Territories. The plan consisted of a simple union called the Sword Border Army. The Sword Border Army was the first United Nations and was an integrated force consisting of the Greyhounds, Indians, and other ethnic groups living in the Territories.

2. **Explain the Historical Treaty, Alliances, or the Negotiation with Native Tribes During the Monarch's Rule**

The Romans were very adept at making treaties with foreign nations based on their interpretation of the situation. The treaty required each individual that participated in observing the peace, and if they attacked any member of the agreement, they were to be treated as if they had attacked a Roman.

3. **List the Strengths and Weaknesses of the Monarch's Military Forces**

    The Roman army was very adept at field tactics and could travel the length and breadth of the lands very quickly (e.g., the legionaries). They can also adapt to any environment that was encountered during the campaign. The Roman army used discipline as a weapon that can snowball into victory at the opportune moment. Roman forces (legionaries) were very organized and were able to fight and defeat larger forces (e.g., The Parthian horsemen). The Romans were very adept at rearranging their ranks and phalanxes. The only weakness of the Roman army was the lack of horsemen (e.g., the Parthians). The Romans were ill-prepared to fight the Parthian horsemen.

4. **Where Was the Monarch's Military Command Center (If Territorial)?**

The Romans had the most organized military command post in all of Europe. The central Roman authority was in the capital of the Roman Empire. The Roman central authority was very centralized and had strict guidelines. Any Roman warship had to have a military commander and a support staff. An example of an early surrender is the Battle of Arsanias when Arsanias admitted defeat by lowering his banner. The Battle of Gaugamela was the best example of coups and guerilla tactics.

5. What Made the Monarch's Regime So Successful?

The Roman Empire was so successful because of the logical foundation and the strictly organized governing body. The Romans had an organized form of taxation, and they had an organized army. The state provided for the military of the Roman army, and the state officials were usually of Roman birth.

6. Describe the Monarch's Form of Government

The Roman Empire was the most successful empire in world history. The Roman military controlled the territory, and the government controlled the military. The governor hired and paid the military.

7. Would You Consider the Monarch's Army to Have Been an Effective One?

The military of the Roman Empire was an extremely effective force. It was one of the largest armies in history, and it had ruthless and highly organized. It could only be broken if their supply lines were severed or if the Romans were overwhelmed by their enemy.

8. What Were the Consequences of the Monarch's Actions on the Native Tribes?

The consequence of the Roman influence was the full integration of the tribes into the Roman Empire. The Romans were adept at assimilating the people they conquered.

The Indians were almost fully assimilated under the rule of the Romans.

9. Compare the Various Benefits and Drawbacks of the Monarch's Reign Over the Greyhounds

The Roman ruler had many benefits of ruling the Greyhounds. The Roman army was on a steep learning curve that eventually resulted in military supremacy. The Romans suffered many drawbacks in their rise to dominance. The Romans initially suffered a military setback. Rome also had to fight against the Slingers and the Scouts.

Setting up the Board

Configure a chessboard successfully without breaking a sweat of the chess is a tough and fun sport. But before you could play chess, you ought to recognize the way to set the board for the proper match.

Chessboard level

The main aspect you have to do is make sure your page is ready up correctly. With a clean board in front of you, there ought to be a white area within the proper bottom nook, irrespective of which side you're on (in white or black portions). At first glance, this can appear like a choice, but the role of the chessboard is one of the maximum crucial components to ensure the chess boards are within the exact region.

Put the Portions

Rooks are elements that appear to be a fort or a cup. The knights look like horses. Bishops are portions with a "frown" groove on top. The king is the most noteworthy piece and has a cross above it. The queen is the second tallest piece with a "crown" above it.

For your rival (darkish sides right now), the setting is sort of the equivalent. The foremost distinction is that while you work from left to right with darkish pieces, first comes the king, at that point, the sovereign.

You will analyze to test this in just one second quickly. Of course, for both sides, the second row at the lowest is full of pedestrians—they are the shortest piece on the board.

As referred to earlier, chess is a traditional strategy and approach suit. It is a struggling game that is performed on a 64 square board with opportunity colors. These squares are referred to as bright and dark squares. The portions are usually black, and the white actions first at the conventional beginning. Each element has 16 portions, as follows:

- Eight infantry (soldiers)

- Two rooks (fortress)

- Two bishops (priests)

- Two knights (guard)

- A king and a queen (Royalty)

To alter the chessboard, set it up so that part of the board moves before you with 8 squares to one site and to the right.

Configure your gamers as follows:

In the bottom row (close to you):

- Put a Rook at each end.

- Going down the middle, put a bishop after every rook.

- Now place one night close to each bishop.

- Spot the king at the rectangular to one facet.

- Put your queen in the ultimate rectangular.

In the front row of the rooks, bishops, knights, and 8 pioneers are positioned one on each rectangular, from left to right. The undertaking of the infantry is to shield the king at any cost.

A technique to remember where the king and queen are to put the lord on the opposite area of shading. If the king is white, he might be in a dark area. If it's black, visit the mild area. It's identical in every case.

All pieces pass in a certain way during the game. Here is a summary of the mobility of each piece:

The snare can pass any wide variety of squares vertically or horizontally yet cannot avoid distinctive components.

Bishops can move any quantity of squares corner to the nook, yet they can avoid exclusive parts.

Knights can pass two squares on a level plane or vertically, and afterward, they can pass one square in any form into an "L" form. The knight is the principal piece that can leap on the board.

The king can pass a rectangle in the direction of any path at any time. The important exemption is the "château" development, where the rook can flow two squares to the ruler, and the lord will circulate two squares in the opposite direction. The final results are that rook and king are on inverse sides of one each other. It has a tendency to be completed using both aspects of the board, as long as neither the euro nor the lord has moved before.

The queen can pass any wide variety of squares in the direction of any route she needs; however, she can't hop anyplace.

Pedestrians can flow one or two squares in their first flow, then one rectangular in every subsequent step.

When nicely configured, your "military" will face every other on the battlefield. The cause of this game is to seize or kill another player's king, the first individual to win the game.

Check, Checkmate, and Draws

In a chess recreation, you ought to take a look at the opposite king. It will help you practice the elements because of the loading agent.

Thanks to this guide, you may enhance your vision at the chessboard, and you may discover spouses faster.

Anastasia's spouse is a general manner to test. The officer took his name from the radical Anastasia und das Schachspiel through Johann Jakob Wilhelm Heinz. It is executed the usage of a knight and a crusher to trap and take a look at the black king.

Andersen's Colleague

Andersen's wife is a well-known manner to test and is known as Adolf Anderson. This officer uses a white rook or a queen to check the black king.

The boot is supported using a pedestrian or a bishop. Anderson's wife is frequently visible before, and very little can be finished to save her.

Arab Neighbor

The wife of Saudi Arabia is a general way to test. The verification officer works with the knight so that the king's diagonal squares are black, so he can capture him with a rook to prepare the check. This dealer can test the rank or report.

The Player behind the Rank

The back pair is a fashionable manner to test. It takes place that a collar or queen controls a king who is blocked through his thick (mainly pedestrian) pieces inside the first or eighth row, and there is virtually no manner to carry the attacking piece to the troubled king. For example, the Black queen cannot capture the white rook.

Bishop and Knight of the Maidana

The bishop and knight officer is a well-known way to check. It takes place that the king and his thick pieces pressure the bishop and knight, the king of the adversary, at the nook of the photo that the bishop can manipulate so that he can deliver his spouse. It is also possible to urgently use the missing king in the deadlock, in which he can be checked.

However, of the four critical missions, along with Queen Kate, Box Matt, and King, and the two suspicious purposes of the bishops, that is a part of the husband's hardest strengths, as he can play 34 full-sport games. Sometimes the result is a draw.

Blackburn's Wife

Blackburn's spouse is known as Joseph Henry Blackburn, and it is an unusual way to check. Using a black liquor (in place of a bishop or queen), the Czech officer restricts the getaway of the black king in square f8. One of the bishops restricts the motion of the black king by using lengthy operating distances, while the knight and bishop work nearby. Threats of Blackburn's spouse may be used to weaken Black's function.

Matt Field

Matt Box is a part of the four most important missions with the queen's spouse, the king, and the 2 bishops and knights. It occurs like the king's side and the empty king's pavement container to the corner or the aspect of the plank.

Corner Couple

A corner couple is a popular manner to test. It is executed by locking the king in a nook using a rook and a queen and using a knight to lease the managing officer.

Cosio's Wife

Cosio's spouse is a fashionable manner to test. The checkmate is an inverted model of Dovetail's husband. It became named after an observation conducted in 1766 by Carlo Cosio.

Similar to Bishop Damiano

Bishop Damiano is an incredible direction to verify. The verification officer makes use of the queen and the bishop, wherein the bishop is used to support the queen, and the queen is conversant in engaging inside the mission. The checkmate is named after Pedro Damiano.

Damiano's Colleague

Damiano's wife is an excellent way to check one of the oldest. It is executed with the aid of closing the king on the floor and using a queen to begin the concluding blow. This rook can also be a bishop or queen.

Damiano's wife regularly comes by sacrificing a collar in report H, then examines the king with the queen in report H, then goes after his wife. Pedro Damiano first posted the Czech agent in 1512. In Damiano's publications, he did not put the white king on the board, which led to his failure to enter many chess databases due to the refusal to accept illegal positions.

The Wife of David and Goliath

David and Goliath's wife is a fashionable way to test. Although David and Goliath's husband can take quite a few forms, they are generally known as the husband in which the infantry is the closing attacking piece, and the enemy's infantry is positioned nearby.

Same Color as Double Bishop

Binary doubling is an ideal way to test. It's like being a wife, but a little easier. The inspector includes the assault on the king, the usage of two bishops, and, consequently, the king is located behind a black pedestal that has now not been moved.

Columbofil Colleague

Dovetail's spouse is a preferred way to check. It includes overthrowing the darkish king in the version displayed on the right-hand facet. It doesn't make a difference how the sovereign is upheld; it doesn't make a distinction which of the other pieces is dark as a knight.

Epaulette's Spouse

Epaulet, or Epaulet mate, by way of its genuine definition, is an inspection agent in which parallel retreat squares are drawn for a king, occupying parts of it and stopping it from escaping. Epaulette's most everyday spouse consists of the king inside the lower back row, stuck among rooks. The visual resemblance among the rooks and the bulbs, the ornamental portions of the shoulder worn over the army uniforms, give it the name of the sheet.

Greco's Colleague

Greek spouses are a fashionable way to test. This Checkmate agent has been named after the best-regarded catalog of the Italian agent Gioachino Greco. It is finished with the aid of the use of the bishop to govern the black king, the use of the black infantry after which through the queen, to check at the king, transferring him to the aspect of the chessboard.

The Wife in File H

H-report mate is a way to test. The inspector includes using an alley that attacks the black king, which is supported through the bishop. This frequently takes place after the castles of the black king in the position of maids in his kingdom. White normally enters this position after a chain of sacrifices in case h.

Pair Rook

The pair of rooks consists of a white rook, knight, and infantry in conjunction with a black infantry to limit the escape of the black king. The knight protects the rook, and the infantry protects the knight.

The Assignment of the King and the 2 Bishops

The king's venture and the 2 bishops are one of the four essential functions alongside the queen's spouse, Matt's container, and the bishop and knight.

It happens that the king, with bishops, forces the bare king at the nook of the image to pressure a likely wife.

The King and the 2 Knights Have a Project

In a game of two knights, the king and the 2 knights cannot force an empty king to be arrested. If the empty king is gambling correctly, this last game has to be drawn. A player makes a mistake most effective if the participant with the empty king is wrong or has already been inside the corner of the board.

Lolli's Neighbor

Lolli's wife is a fashionable manner to check. The summary consists of the infiltration of Black's fiancé's position using his leg and queen.

The queen normally arrives in rectangular h6, the usage of the sacrifices in report H, dubbed after Giambattista Lolli.

A Colleague of Max Lange

Max Lange's spouse is a general manner to test. The Czech officer is appointed Max Lange. It is done by the use of the bishop and queen to manipulate the king.

Murphy's Mate

Murphy's spouse is a well-known manner to check. Named after Paul Murphy, this is achieved by the use of a bishop to assault the black king and a rook and a black and white pedestrian to fasten him up. In many ways, he's very much like Corner's wife.

Opera Colleague

Mate Opera is a widespread manner to check. It works through attacking the king within the returned with a rook using a bishop to protect him. A pedestal or other piece apart from the knight of the enemy king is used to restrict his movement.

This teammate became named the Opera after the interpretation of Paul Morphy in 1858 in a Paris opera against Dunk Carl de Brunswick and Count Isouard.

Comrade Pillsbury

Pillsbury's wife is a preferred verifier, named after Harry Nelson Pillsbury. As shown on the right, it works with the aid of attacking the king or a pier or bishop. The king can be in g8 or h8 during checkmate.

The Spouse of the Queen

The queen is one of the four vital missions along with Boxing, King and two Bishops, and Ismail and Knight.

It happens when the party with the queen and king forces the naked queen to the brink or corner of the council. The queen completely examines the naked king, and the pleasant king supports her.

Reti's Mate

Reti marriage is a popular way to test. The Czech officer is known as Richard Reti. Do this by grabbing the enemy king with four portions, which are inside the flying fields, after which attacking him with a bishop who's covered by a rook or a queen.

Colleague Mate

Intimate pairing is a well-known way to test. It occurs that a knight controls the kingdom, which is suppressed (besieged) by its thick portions, and he has nowhere to move, and there is virtually no manner to defeat the knight.

The Neighbor's Sleep

Suffocation is a trendy way of checking. It is performed by the usage of the knights to assault the rival king and the bishop to restrict the king's getaway routes.

Swallow's Tail

Swallow's tail, also referred to as Guéridon's wife, is a standard manner to check. It works by attacking the enemy king with a queen included using a rook. The rooks of the enemy king block his escape device. He is just like Epaulette's spouse.

Chapter 2. The Goal in Chess

The chessboard has 64 squares with alternating colors. Each opponent will have 16 pieces divided as follows: 8 pawns, 2 knights, 2 bishops, 2 rooks, 1 queen, and 1 king. Each player will have a set of 16 chess pieces in the same color.

Before a game is begun, the pieces need to be laid out in a specific fashion on the chessboard each time a new game is played. Whoever gets the white pieces gets the first move. Each opponent would need to place their pieces on the chessboard side nearest to them. The second row nearest each player is filled with pawns. On the first row nearest to each player is the other pieces in a specific formation. The black queen is on the center of the first row, on the black square; the opposite is true for the white queen—it goes on the middle white square of the first row. The two rooks go on each opposite end of the first row, followed by the knights on each side, then the bishops, and the king and queen. Once again, remember that the queen goes to its matching-colored square.

Once the chessboard is set, the game can begin with the player owning the white piece moving first. And then the opponent moves, followed by the first player, and so on and so forth.

The Goal of the Game

Of course, like in any game or sport, the goal is to win the game. Winning a chess game only means one thing—checkmate. Checkmate means when you have cornered the opponent's king in such a way that whatever movement the opponent's king makes, it will be captured. All other chess game endings lead to a draw.

Getting to Know a Few Moves

There are some specialized moves in the game of chess and some of which we have already discussed earlier, such as promotion and en passant.

Capturing

When capturing an opponent's piece, move your piece according to that specific piece's type of movement, and replace your opponent's piece with your triumphant piece and not a space more. Remember that only horses or knights can jump over other chess pieces.

Castling

Castling is an important chess movement that helps to protect your king. Since at the beginning of the game, the center of the chessboard is where the battle ensues, and since your king is situated in the center edge of the board, it is definitely a good idea to place it at the corner of the board where it can be defended by a rook. This is done easily and with just one movement through castling. Further, castling can only be done between a king and your rook and no other chess piece.

Before you can make a castling move, there are some parameters that you need to meet first, such as there should be no other chess pieces in between your king and rook; the rook that you are going to use for castling should not have moved since the beginning of the game, this also goes for the king—if either or both of them has already moved then castling is no longer an option; if the king is in check, it cannot move out of it through castling; a king cannot castle if it is going to cross a square that is being attacked by an opponent's piece; and lastly, a king cannot castle if it places itself in a check position.

Therefore, castling should be done as soon as possible, or your opponent can say that you have already moved your rook or king, thereby disqualifying you from making a castling move.

So, without further ado, let me teach you how to do castling. There are two ways to do castling, and these are:

Castling Kingside

When castling, the king is moved only two squares. So, when castling kingside, this means moving the king to the rook nearest its side—hence the name castling kingside. And then the rook is now moved on the opposite side of the king with no squares in between them. When castling kingside, the king is nearer the corner of one board with just one more square, and on the other side is the rook. This makes the king more protected with pawns in front of it. Of the two castling moves available, castling the kingside way provides more or better protection and should be your first choice.

Castling Queenside

When castling queenside, this only means that you are moving your king against the rook that's nearest to where your queen used to sit. So, the same castling rule still applies: the king should be moved only two spaces towards the rook, and the rook is moved on the king's opposite side. With this type of castling, the king is two spaces away from the chessboard's corner.

Check and Checkmate

Check is different from checkmate. Check happens when a king is being attacked by an opponent's piece, then that is called check. With a check, there will be three scenarios to deflect a check move. One, the player with a check king can just move the king to a safer square. Two, the player with the checked king can move a piece that can capture the opponent's piece that has checked his or her king. And the third possible move is to block the check by placing another chess piece in between the king and the opponent's piece. So, getting your king checked does not mean the end of the game, but just that your king is in peril of being capture.

Checkmate, on the other hand, signals the end of the game, and the player who checkmates their opponent's king wins. So, what happens during a checkmate? During a checkmate, the king is placed in a predicament that no matter where it moves, there is an opponent's piece already waiting to capture it if it moves on that spot. Remember that although the king's movement can be in any direction, it is crippled by the fact that it can only move one square at a time. Thus, checkmating a king is possible but not easy.

Once a player cannot do any of the following during his or her turn, then she or he is checkmated: Capture the attacking piece, move a piece in between the king and the enemy piece to block the check, and move to a square that is not being attacked by the enemy piece.

Basic Chess Rules

We have already covered a few chess rules when talking about how to capture opponent chessmen, 'en passant,' promotion, and movement of each chessman. However, there are more basic rules that you need to know in order to play a correct game of chess. So, here are some of the basic rules that you should know about:

- Each game should be done in a respectful and friendly manner. Do not do anything to annoy, irritate, or distract your opponent. This is important because, as we have said before, chess is a mental game.

- There are chess games where a clock is used. If and when using a clock, the clock's button should be pressed by the hand that moved the chessman on the board and not by your other hand.

- When castling, the first piece to be moved should be the king, and the rook should be placed at the opposite side of the king.

- An individual can adjust a chessman on the board by saying "j'adoube," which means 'I adjust' in French.

- Once you touch a chessman on the board, you need to move that piece unless it places your king in peril. This is known as the 'touch-move rule.'

- Make your move using only one hand.

- A pawn upon being promoted, its powers will be immediate even if this means that it checks or checkmates the opponent's king.

- To play, the chessboard should be placed in such a way that the first white square is located on the right side of the player.

20 Strategies for Achieving the Goal

When starting a chess game, I look out for three things. Yes, another ABC list to remember! Since I have been a chess instructor these past seven years, I continue to find new and better methods, such as these, that help my students remember concepts more effectively.

- Activate your pieces
- Bring your king into safety
- Control the center

Activate your pieces: Bring your knights and bishops out! Don't move only Pawns.

Bring your king into safety: Castle your king. Get your king away from the center, where the enemy pieces are waiting to attack.

Control the center: Control one of the center squares or get your pieces close by.

1. Control the Board (Especially the Center)

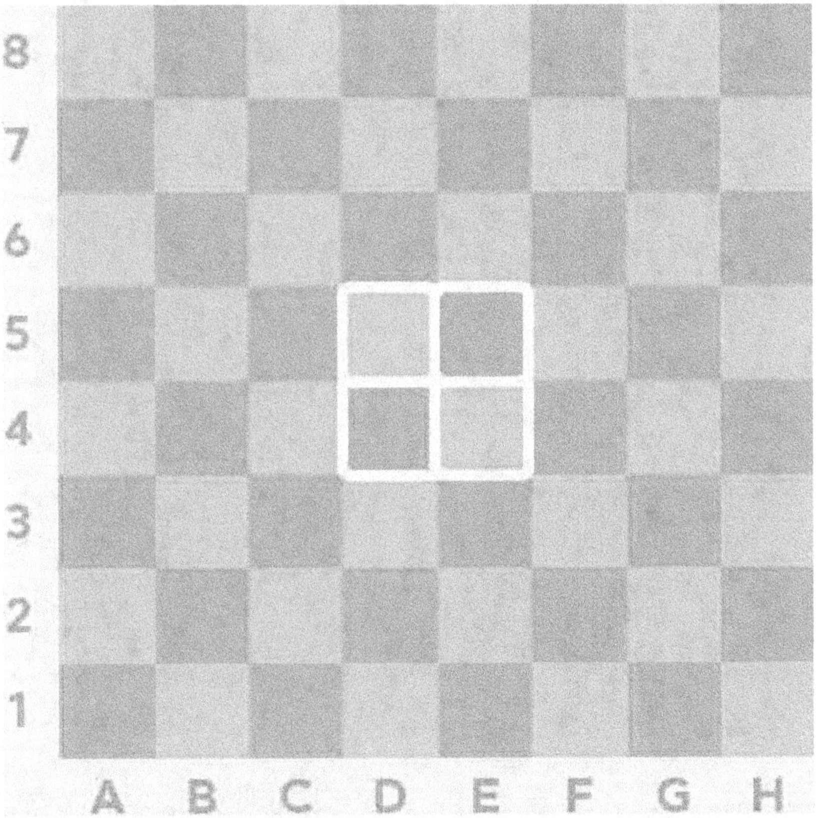

The more pieces you have in the center, the more control you have of the board. When a piece gets developed in the center, it has more mobility, whereas a piece on the side of the board usually has limited movement.

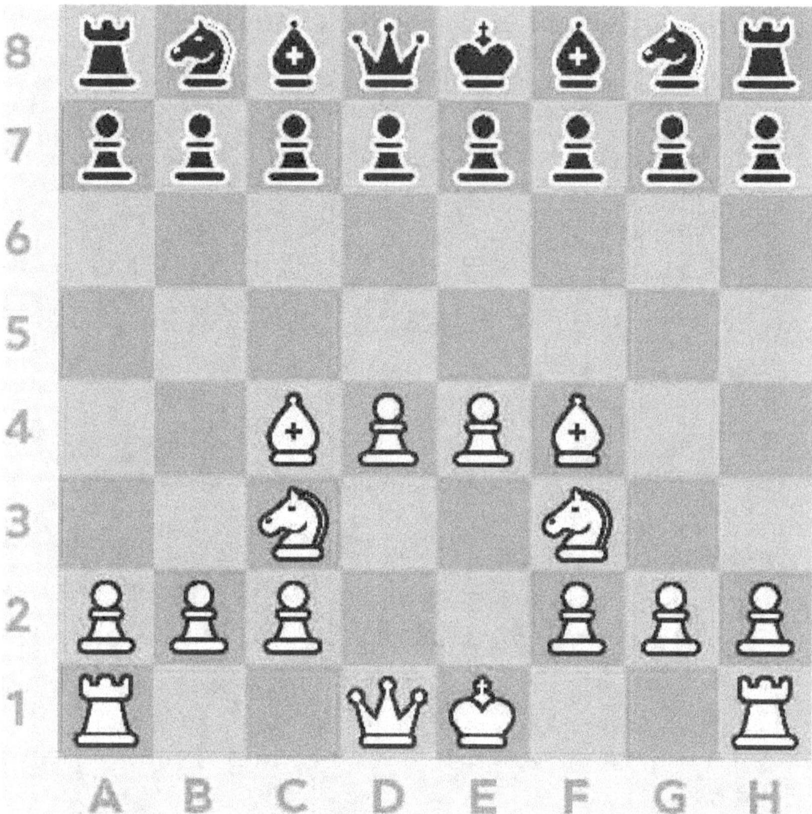

This is a perfect position White (or Black) would like to achieve. This diagram, however, is not realistic since it doesn't seem like Black has made any moves yet. I am showing the dream position for every chess player— offering control of the center with all one's pieces. White's next move would be to castle and then move the queen to e2 or d2 to link the rooks.

What is a chess opening? A chess opening is a series of moves at the beginning of the game. The opening usually lasts for 10 to 15 moves, and then the middle game starts. There are dozens of different openings and more than 100 variants of each one. Variants are the different moves that can be played after the starting moves. For example, 1. e4 c5 is called the Sicilian Defense opening. All moves following this are called chess variants. 1. e4 c5; 2. Nf3 g6 is now called Sicilian Defense: Hyperaccelerated Dragon variant. For each variant, there's a code.

I wouldn't worry too much about openings until you are fairly familiar with the rules of the game. The following diagrams illustrate two common openings.

Four Knights Game: Italian Variation C50

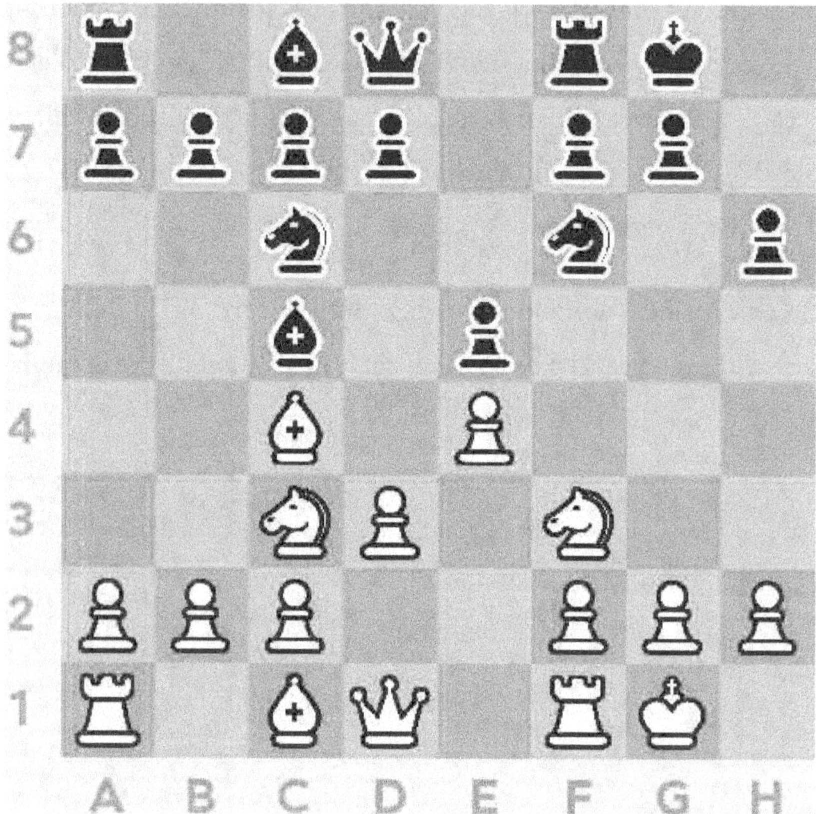

1. e4 e5 White Pawn to e4, Black Pawn to e5. Both Pawns protect the center.

2. Nf3 Nc6 White knight to f3, Black knight to c6. Develop the knights toward the center, where they control more squares, unlike their positions at the sides of the board.

3. Bc4 Bc5 White bishop to c4, Black bishop to c5. Develop the bishops to protect the center and surrounding squares. The bishops observe each other's weaknesses, the f Pawns, but they also open the way for their kings to castle.

4. 0-0 Nf6 kingside castle, Black knight to f6. Castle the king. Black will castle on the following move because they must develop their knight first (which also attacks the e4 Pawn).

5. Nc3 0-0 White knight to c3, kingside castle. White develops its knight and protects the e4 Pawn. Both sides now control the center and have castled their kings. Next, they must develop their bishops, but the Pawns are blocking the way.

6. d3 h6 White Pawn to d3, Black Pawn to h6. Black stops white from developing the bishop to g5 while creating a fleeing square for their king in the future.

Queen's Pawn Game: Chigorin Variation D02

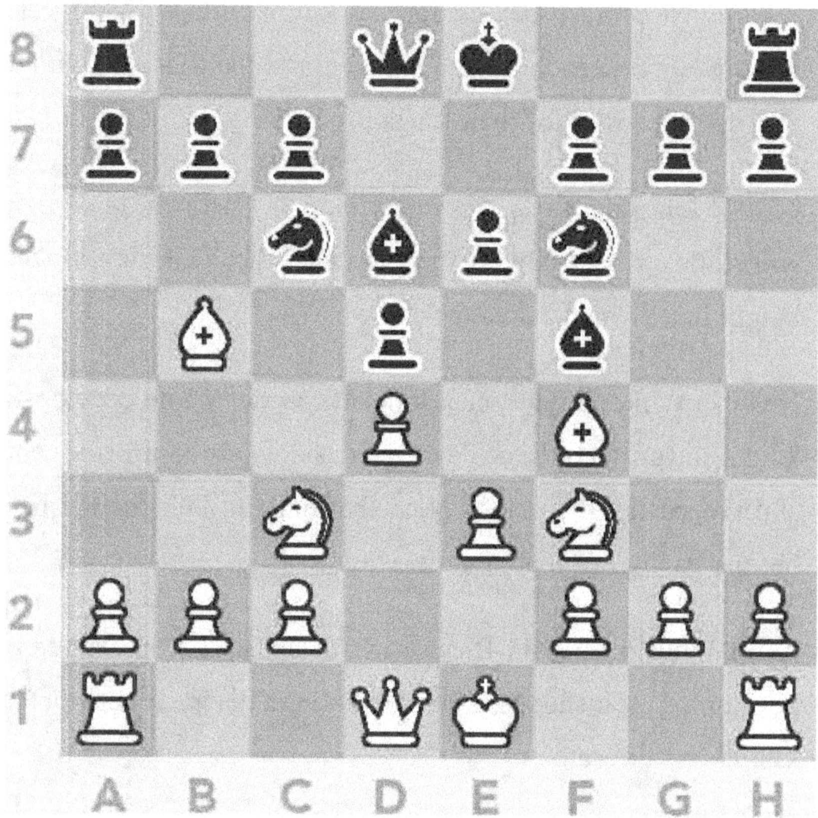

1. d4 d5 White Pawn to d4, Black Pawn to d5

2. Nf3 Nc6 White knight to f3, Black knight to c6

3. Bf4 Nf6 White bishop to f4, Black knight to f6

4. Nc3 Bf5 White knight to c3, Black bishop to f5

5. e3 e6 White Pawn to e3, Black Pawn to e6

6. Bb5 Bd6 White bishop to b5, Black bishop to d6

At this point in the game, most of the white's and black's pieces are out. On the next move, both sides are ready to castle their kings.

2. Don't Give Up Your Pieces Easily

One of the most important concepts for beginners in chess is understanding the values of the different pieces—this allows you to calculate when to capture pieces and when to hold back. After all, you don't want to give your opponent unnecessary gifts—this will only help them beat you!

Piece	Points
King	Game
Queen	9
Rook	5

Bishop	3
Knight	3
Pawn	1

Hanging Pieces

The term hanging in chess means unprotected. Therefore, an unprotected piece is known as a hanging piece.

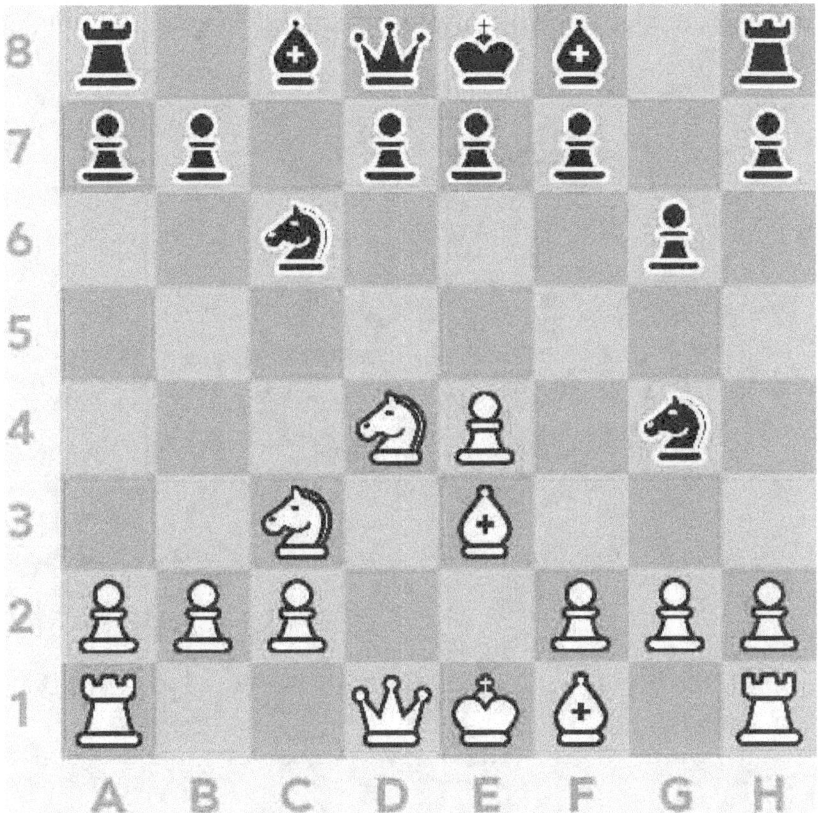

Black just moved their knight to g4 in hopes of capturing the bishop on e3. The problem is the black knight moved onto a square where it's not protected by one of its own. The white queen can simply capture the knight for free: Qxg4.

Capturing Protected Pieces

When capturing protected pieces, a little bit of mental math is involved.

Piece you captured – piece you gave up = equal, losing, or winning points.

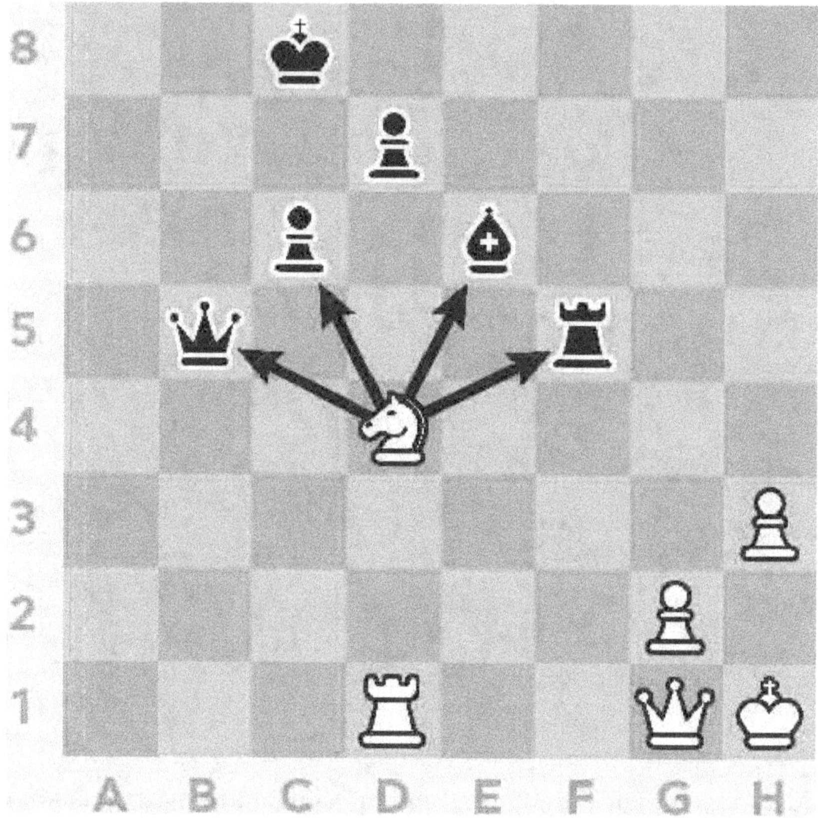

Both sides possess the same amount of points. It's White's turn to move, and the knight is attacking four pieces at once, but they are all protected. Which one should the knight capture?

1. Nxb5 the queen is worth 9 points. If the knight captures the queen, White will lose their knight because of the Pawn or Rook. Math: 9 – 3 = 6. White wins 6 points and the strongest piece!

2. Nxc6 the Pawn is worth 1 point. If the knight captures the Pawn, White will lose their knight because of the Pawn. Math: $1 - 3 = -2$ points. White loses 2 points for capturing a weaker piece.
3. Nxe6 the bishop is worth 3 points. If the knight captures the bishop, White will lose their knight because of the Pawn. Math: $3 - 3 = 0$. This is called a trade (equal). No one wins or loses points.
4. Nxf5 the rook is worth 5 points. If the knight captures the rook, White will lose their knight because of the bishop. Math: $5 - 3 = 2$. White wins 2 points for capturing a stronger piece.

Answer: The best capture for the knight is the queen on b5, which will make him win 6 points.

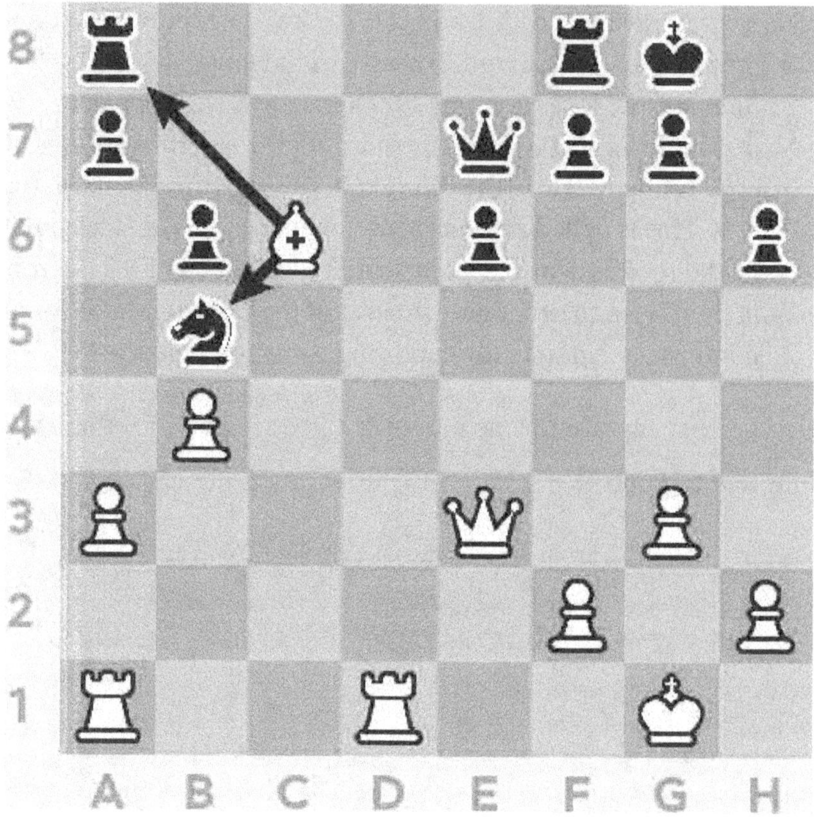

White is down 1 point. It's White's turn to move, and the bishop is attacking two pieces at the same time: the protected rook and hanging knight.

1. Bxa8 the rook is worth 5 points. If the bishop captures the rook, White will lose their bishop because of the other rook on f8. Math: 5 –3 = 2. White wins 2 points for capturing a stronger (but protected) piece.

2. Bxb5 the knight is worth 3 points. If the bishop captures the knight, White won't lose any points since the knight is not protected by any of Black's pieces. Math: 3 − 0 = 3. White wins 3 points for capturing an unprotected piece.

Trading Pieces

A trade takes place when a piece of the same value captures a protected piece of the same value; for example, queen for queen, knight for bishop, or Pawn for Pawn.

Reasons for trades:

- When you are ahead in material, to remove any of your opponent's pieces that can potentially counterattack your plans

- When there are open files/ranks/diagonals

- To remove a piece defending certain squares

When you decide to trade a piece, make sure you don't give up a strong one (with more mobility) for a weak one (with less mobility).

Exceptions

Many games in chess include opportunities for sacrifice. A sacrifice is when you give up material to checkmate or gain more at the end.

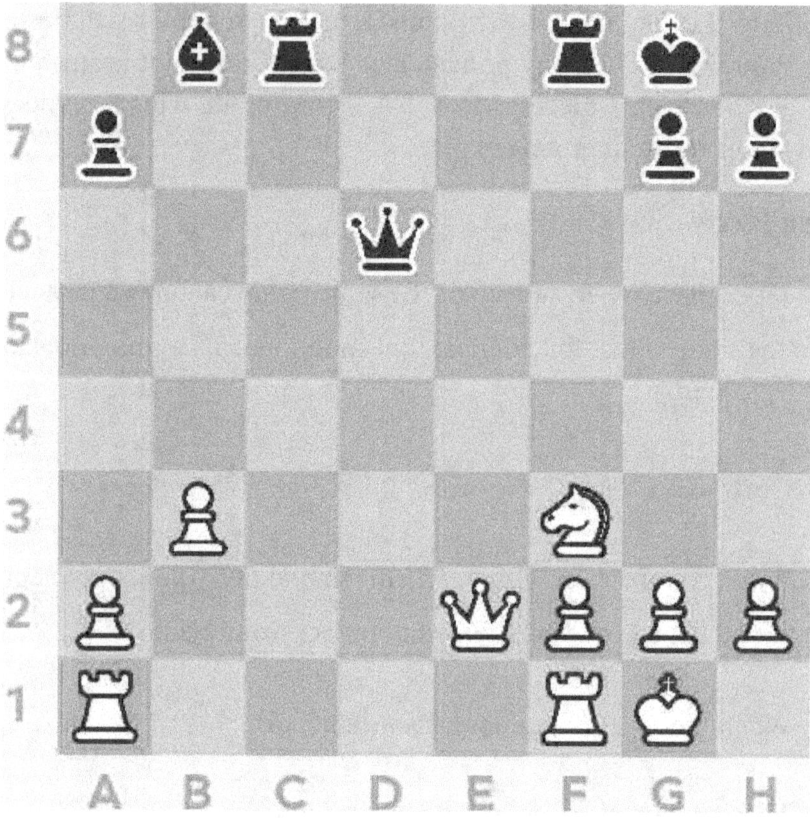

Black is down two Pawns but still maintains a strong position. The queen and the bishop are looking at the h2 Pawn, which is doubly protected by the king and the knight.

1. Rxf3! Black rook captures knight on f3. Black captures the protected knight on f3! Why on earth would Black give up a 5-point piece for a 3-point piece? The knight is protecting checkmate on h2. This is also called removing the guard or removing the

defender. White's best move is to play 2. g3 to protect against the threat. But then Black will respond with 2. Rd3, getting the rook to safety (i.e., the rook will be protected by the queen) and a winning advantage for capturing White's knight for free.

2. Qxf3?? Qxh2# White queen captures knight on f3 (terrible move). Black queen captures Pawn on h2, checkmate. White is checkmated. The king can't run away to h1 because it's protected by the queen. The king can't capture the queen because it's protected by the bishop. There is no blocking possible here since the queen is too close to the king. This is a good example of how every move requires you to think ahead to its consequences. In this case, capturing the rook seemed like a very straightforward idea but turned out to have a negative outcome.

3. Ask Yourself Why

Whenever I play a game of chess, I always ask myself why my opponent made a move. It doesn't matter if I'm playing against a weaker or stronger player. Weak players can still make occasional strong moves, and strong players can make mistakes. That's what makes us human.

When your opponent makes a move, ask yourself these three questions:

1. Is the piece that just moved attacking anything?

2. Are any of the other pieces attacking mine?

3. Did the piece that just moved stop protecting another piece or stop protecting against a threat?

Move: Attack

If your opponent made a move, and your piece is now under attack, first see if you can capture it with an equal-value or lower-value piece. If no capture is possible, move it away unless it's the same or higher value as the attacker; in that case, you can also protect it. If the attacker is a lower-value piece, you most likely want to move your piece.

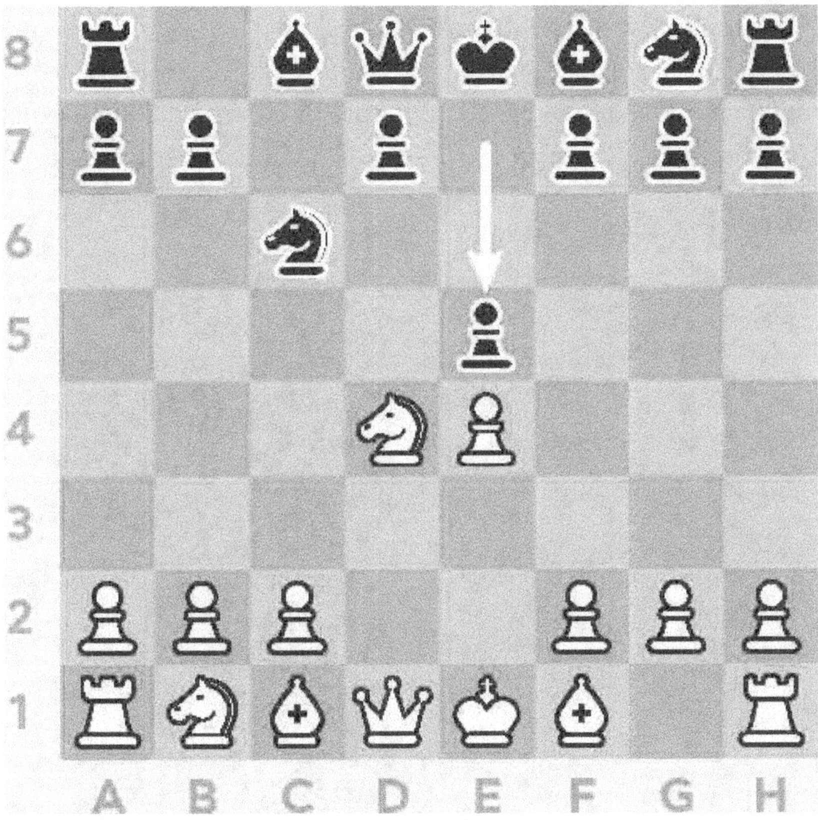

- **Why did Black make this move? Are they attacking anything?**

Black is attacking the knight. White can't simply protect their knight because the attacker is a lower-ranked piece. If 1. Be3 exd4, then White losing a knight for a Pawn would equal −2 points. We must do something about the knight.

• What should White do now?

The knight can move to b3, b5, e2, f3, or f5, or capture the knight on c6 (a trade). It can't leave the knight on the square or move it to e6 because of the d7 and f7 Pawns.

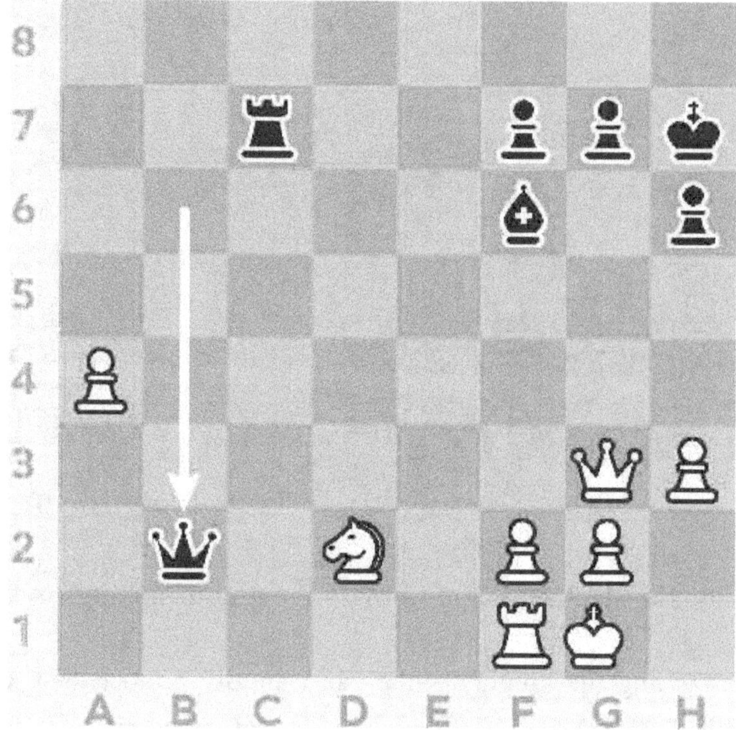

• Why did Black make this move?

Black is attacking the knight. White has multiple possibilities—Ne4, Nf3, and Rd1 to get the knight out of danger—but sees that when the queen made a move, it stopped protecting the rook. Who cares about our 3-point piece if we can capture a 5-point one? 1. Qxc7.

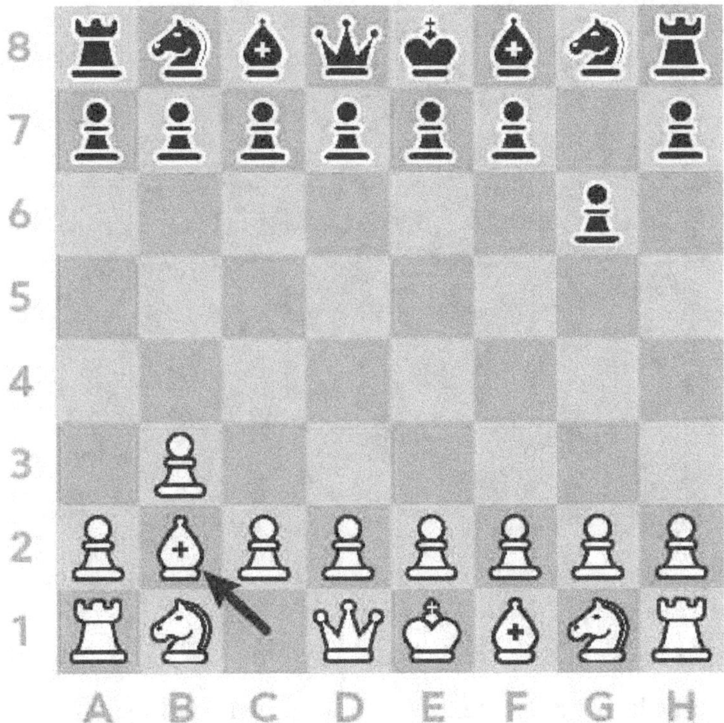

Move: No Attack

If your opponent made a move and none of your pieces are under attack by that piece, see if it was protecting another piece or protecting against a threat. A lot of beginner chess players forget about pieces they were defending earlier.

White moved the knight. Is it attacking anything? No. The Pawn on c6 is protected by the Pawn and queen. Was the knight defending anything? Yes. It used to protect the other knight.

1. Nd4?? 2. Bxd2

Black moved the knight. Is it attacking anything? No. Was it protecting anything? Yes. It used to protect the e8 square, so now White is able to checkmate: 1. Nd5; 2. Re8#.

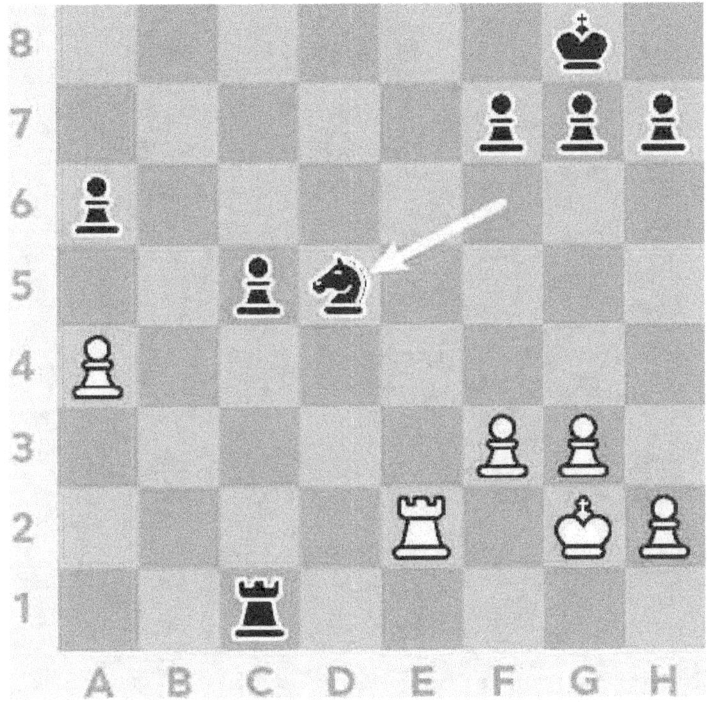

4. Look at the Whole Board

In my opinion, this strategy is critical, and many beginner players fail to consider it. If you scan the whole board before and after your move, you will make fewer mistakes. I have failed to remember this advice myself on certain occasions. During those games, I was so focused on one side of the board that I didn't notice my opponent's attack on the other side.

Scanning the board before and after your move gives you a much better chance of noticing:

- unexpected attacks on your pieces

- potential attacks you can make on your opponent's pieces

- pieces and threats that are no longer protected by your opponent

Yelizaveta Orlova – N.N.

1. b3 g6 White Pawn to b3, Black Pawn to g6

2. Bb2 Nf6? White bishop to b2, Black knight to f6 (bad move)

3. f4 d5 White Pawn to f4, Black Pawn to d5

4. Nf3 c5 White knight to f3, Black Pawn to c5

5. e3 Ne4?? White Pawn to e3, Black knight to e4 (terrible move)

Black's move, at a glance, seems good. The knight is located on a central square; it's guarded by the Pawn on d5, and none of White's pieces can capture it.

When Black looks at the whole board, however, they notice that the bishop on b2 is attacking the rook on h8, which also happens to be unprotected!

Bxh8! White bishop captures rook on h8 (excellent move). If White had only focused on the piece that just moved (and not the entire board), they wouldn't have won 5 points!

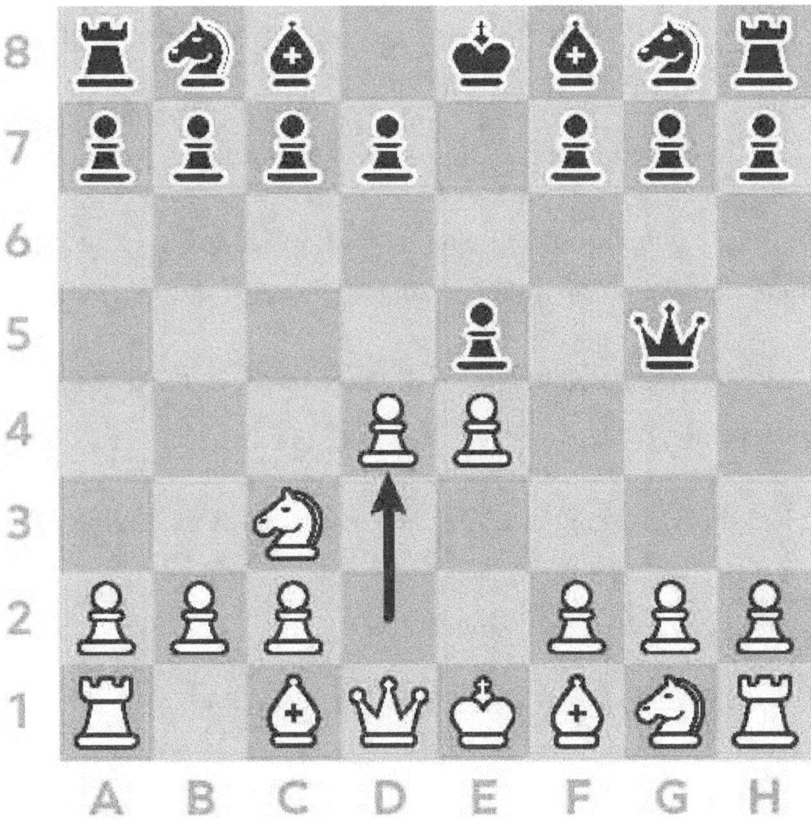

5. Protect Your King

As discussed in the castling section earlier, the king is vulnerable in the center of the board unless it's in the endgame. You wouldn't want your king hanging around your opponent's pieces. Castling, either kingside or queenside, will get your king away from the center. There are rare exceptions when top players don't castle, but in 99 percent of games, castling is involved.

Castle as Quickly as Possible

The longer you keep your king in the center, the more likely it will be in danger. Try to castle in the first 10 moves of the game.

Here is a nice example, called Legal's Mate, which involves keeping the Black king in the center (while the White king is one move away):

1. e4 e5 White Pawn to e4, Black Pawn to e5. Both Pawns control the center.

2. Nf3 d6 White knight to f3, Black Pawn to d6. The White knight attacks the Pawn on e5, and Black defends differently than usual. More common is the move 2. Nc6.

3. Bc4 Bg4 White bishop to c4, Black bishop to g4. White and Black develop their bishops. White's bishop eyes the f7 Pawn, but it's protected by the king. Black's bishop pins White's knight, which can't move because the queen is behind. The queen and Pawn are protecting the knight, which means if the bishop captures the knight, the next move will result in a trade. Black captures a 3-point piece, and White captures a 3-point piece, as well.

4. Nc3 g6? White knight to c3, Black Pawn to g6 (bad move). Black doesn't see the next move coming, but before we get there, let's

analyze this position. White has developed three strong pieces, the two knights and the bishop, while Black has developed only the bishop. Which side is about to castle? White, because the bishop and knight are no longer in the way.

5. Nxe5!! Bxd1?? White knight captures Pawn on e5 (excellent move), Black bishop captures queen on d1 (terrible move). Why would the knight capture the protected Pawn? Not only that, but allow Black to capture their queen? If Black captures the knight, 5. dxe5; 6. Qxg4. White captures both a Pawn and a bishop, while Black has captured only a knight. Therefore, White wins a Pawn—this move is better than Black's, but Black is too greedy and goes for the queen!

6. Bxf7+ Ke7 White bishop captures Pawn on f7, check, Black king to e7. The king can't capture the bishop or move to d7.

7. Nd5# White knight to d5, checkmate. Checkmate! A really beautiful example of keeping the king in the center. Yes, White's king is also not castled yet, but they can do so on any next move.

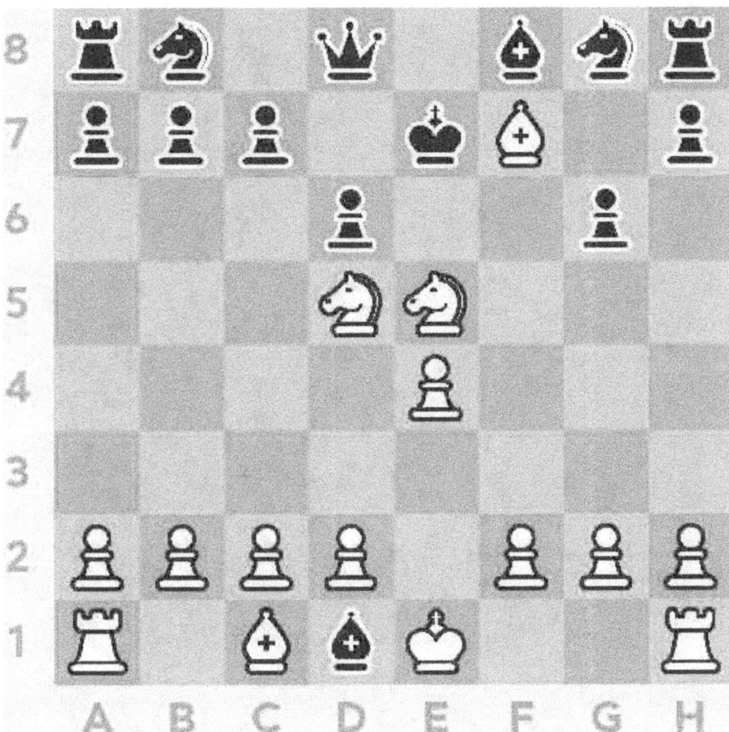

Create a Fleeing Square

When you castle kingside, you should move your h Pawn to give more space to the king. Many players don't create a fleeing point and end up falling for the back-rank checkmate (see Check and Checkmate).

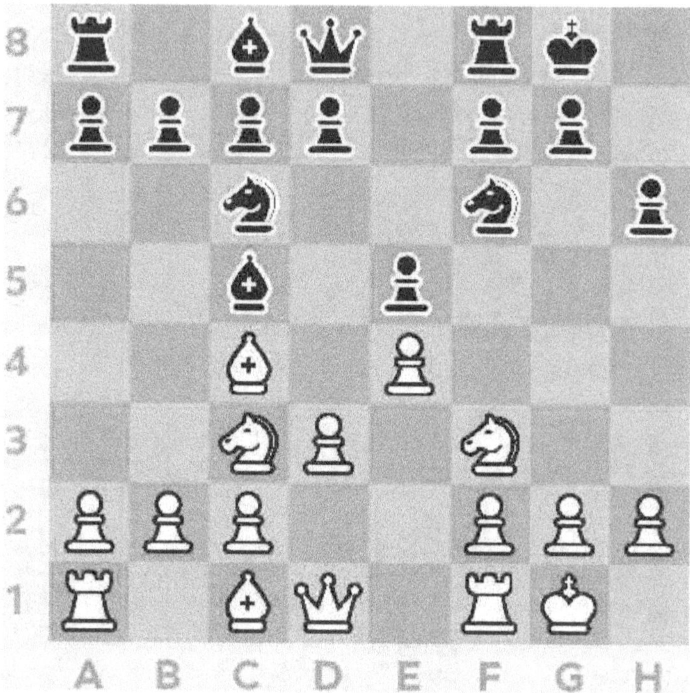

Black's last move was Pawn to h6. This move is to prevent the bishop from moving to g5 but also to let the king breathe and provide the possibility to escape to h7 if needed.

Fianchetto Bishop

A fianchetto bishop is the development of your bishop on the longest diagonal (either a1 to h8 or a8 to h1) while hiding it in a fortress of Pawns. I really like to call the fianchetto bishop "the sniper" since it hides and waits for the right moment to strike! At the very beginning of the game, the bishop doesn't seem particularly strong, but once other pieces are traded off, and diagonals open up, it becomes a monster.

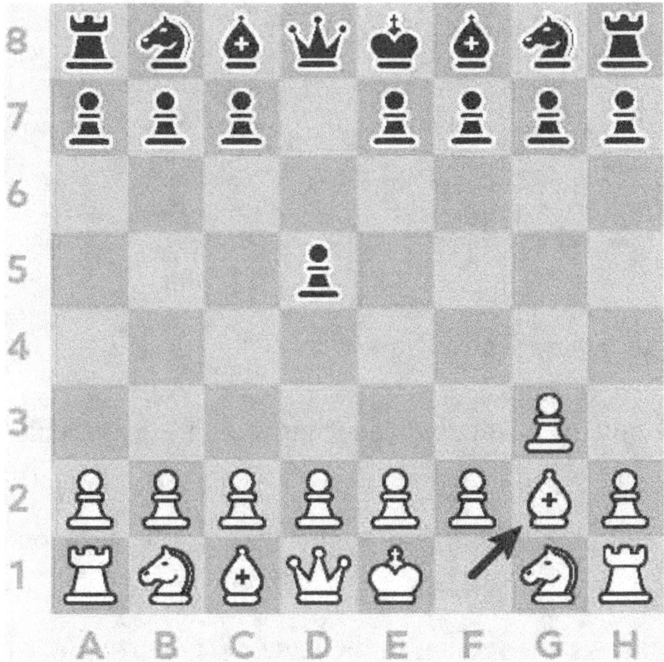

So, why is the fianchetto bishop in this section of the book?

The fianchetto bishop can serve as extra protection for the king, but it can also stop back-rank checkmate. It can stop this by acting as a blockader; once the bishop has moved, the king will have a fleeing square.

Take care not to trade off your fianchetto bishop for your opponent's knight, or you will leave a hole in your king's castle.

6. Use All Your Pieces

Make sure all your pieces are out and ready to fight! If your knight hasn't moved yet, and 10 moves have passed, move it next! Imagine you're playing on a soccer team, but two teammates are just sitting on the field doing nothing. Will the team do well against the opposing team who has all team members participating in the game? Of course not. Similar to soccer, chess requires teamwork! The majority of checkmates don't succeed using just one piece—on the contrary, it takes multiple pieces assisting in the attack.

Emanuel Lasker was a German Grandmaster and 27-time World Champion. Lasker is the only person to hold the World Champion title that long. The following game bearing his name is extremely renowned— today, it's still considered one of the best checkmates!

1. Qxh7 Kxh7 White queen captures Pawn on h7; Black king captures queen on h7. Oh, Lord! White sacrifices their queen for a Pawn. Why? We shall see.

2. Nxf6++ Kh6 White knight captures bishop on f6, double-check, Black king to h6. It's a double check. The king's only response is to move. The knight can't be captured because the bishop is still attacking the king. If 2. Kh8, 3. Ng6# (2. Black king to h8; 3. White knight to g6, checkmate).

3. Ng4+ Kg5 White knight to g4, check, king to g5. Bringing in another knight (second piece) to the attack!

4. h4+ Kf4 White Pawn to h4, check, Black king to f4. Third piece.

5. g3+ Kf3 White Pawn to g3, check, Black king to f3. Fourth piece.

6. Be2+ Kg2 White bishop to e2, check Black king to g2. Luring black more into white's camp.

7. Rh2+ Kg1 White rook to h2, check, Black king to g1. Now the rooks come in.

8. Kd2# White king to d2, checkmate. Beautiful checkmate. White used their pieces as a team to achieve the goal.

7. Think Several Moves Ahead

Thinking several moves ahead is extremely important in chess, and it mostly comes with practice. I did many chess puzzles as a kid: both checkmate problems (in x number of moves) and tactical ones. In my spare time, I read many books on tactics, strategy, and endgame concepts. However, you can't become a stronger player by studying alone—playing chess against others allows you to implement what you have learned.

1. c4 e5 White Pawn to c4, Black Pawn to e5. Both starting moves are great. White protects the d5 center square, and Black protects the d4 center square.

2. e3 Bb4 White Pawn to e3, Black bishop to b4. White's move opens up the way for the bishop. Black develops their bishop but not on the best square, since it's not really accomplishing anything there.

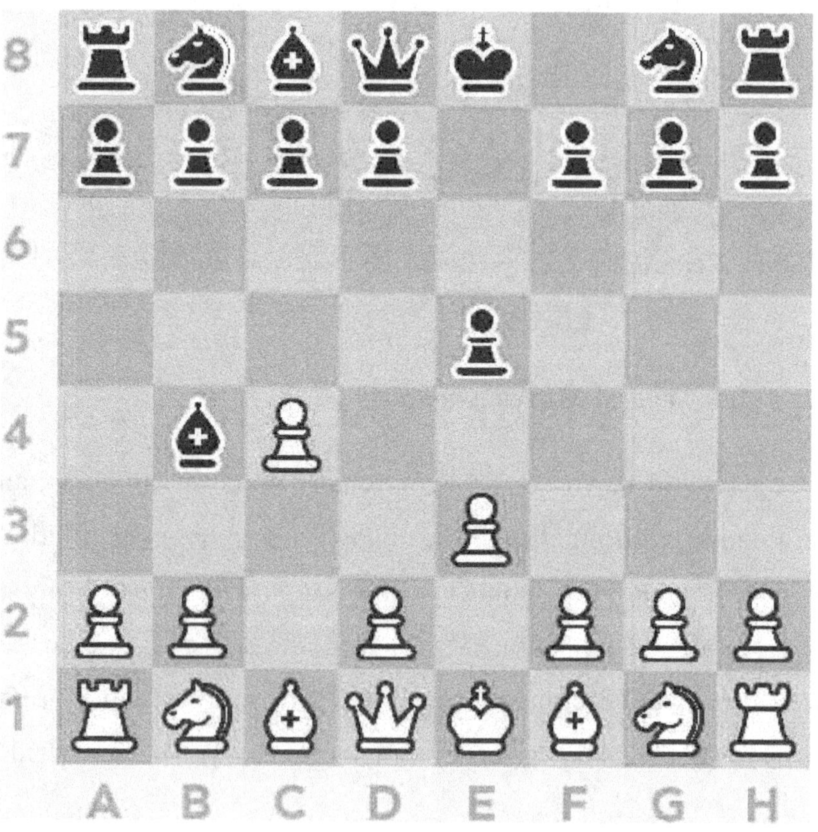

3. a3 Ba5?? White Pawn to a3, bishop to a5 (terrible move). White attacks the bishop and gains a tempo, or gains time. Black loses a tempo because it just developed the bishop and has to move it again instead of another piece. Black's move is a big mistake because they're moving onto a square where the bishop will have limited mobility; after Ba5, the piece has only one square to which it can move. Now, White can trap the bishop in two moves. 3. Bishop to e7 would have been a much better move.

4. b4! Bb6 White Pawn to b4 (good move), Black bishop to b6.

5. c5! White Pawn to c5 (good move). White will win the bishop for a Pawn. Yay!

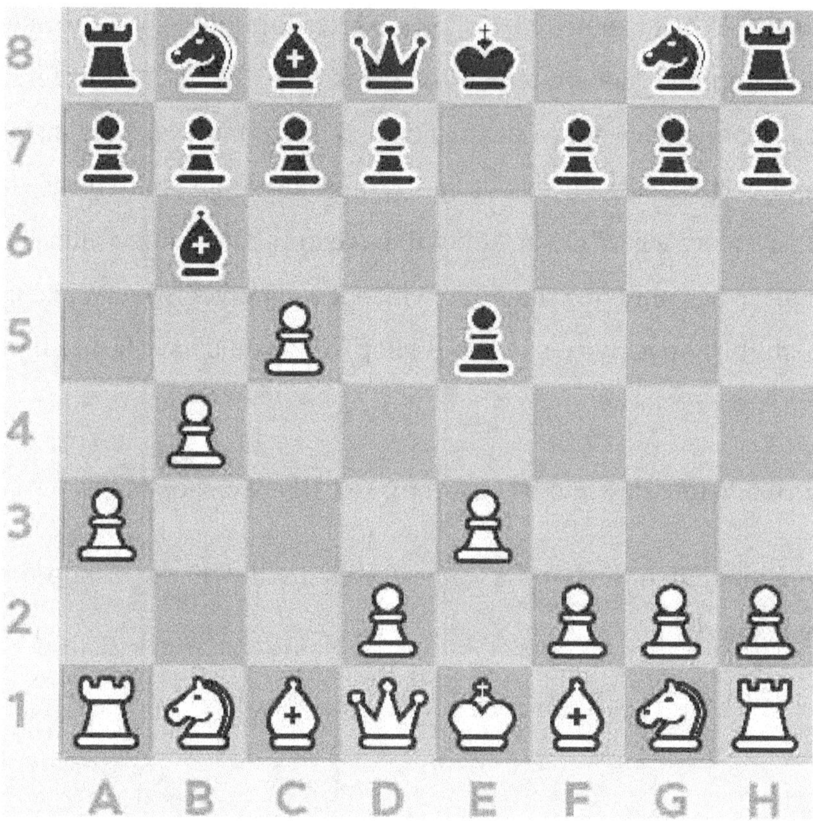

8. Go for a Quick Checkmate

Weakest Pawn

The weakest Pawns in the starting position are the f2 Pawn for White and the f7 Pawn for Black. These are vulnerable because they are the only Pawns (out of eight) protected only by the king. If these Pawns have one attacker and one defender, they're not safe enough to capture. On the other hand, if White or Black manages to add a second attacker, so they have two pieces attacking these Pawns (still with only one defender), they will be easy enough to capture.

Scholar's Mate

Scholar's Mate is the four-move checkmate strategy. (In Russian, it's called Kid's Mate.) If you play against a beginner, I'd suggest trying it out once, but I doubt you will be able to use it a second time against the same opponent!

1. e4 e5 White Pawn to e4, Black Pawn to e5. Good starting moves by both sides.

2. Qh5 Nc6 White queen to h5, Black knight to c6. White's queen is attacking the e5 Pawn (without defenders) and putting pressure on the f7 Pawn (can't capture the Pawn, protected by the king). The Black knight moves to c6 to defend the e5 Pawn. It would be terrible to lose the queen for capturing a protected Pawn, as the queen is worth 9 points, and the Pawn is only worth 1 point.

3. Bc4 Nf6?? White bishop to c4, Black knight to f6 (terrible move). White develops their bishop to c4 to attack the f7 Pawn. The f7 Pawn is now under attack twice (by the queen and bishop), and there is only one defender (the king). Black must protect the Pawn a second time but fails to see the threat—instead, they play Nf6 to attack the queen.

4. Qxf7# White queen captures Pawn on f7, checkmate. Checkmate! The queen is checking the king and protecting square e7. The king can't capture the queen because it's protected by the bishop.

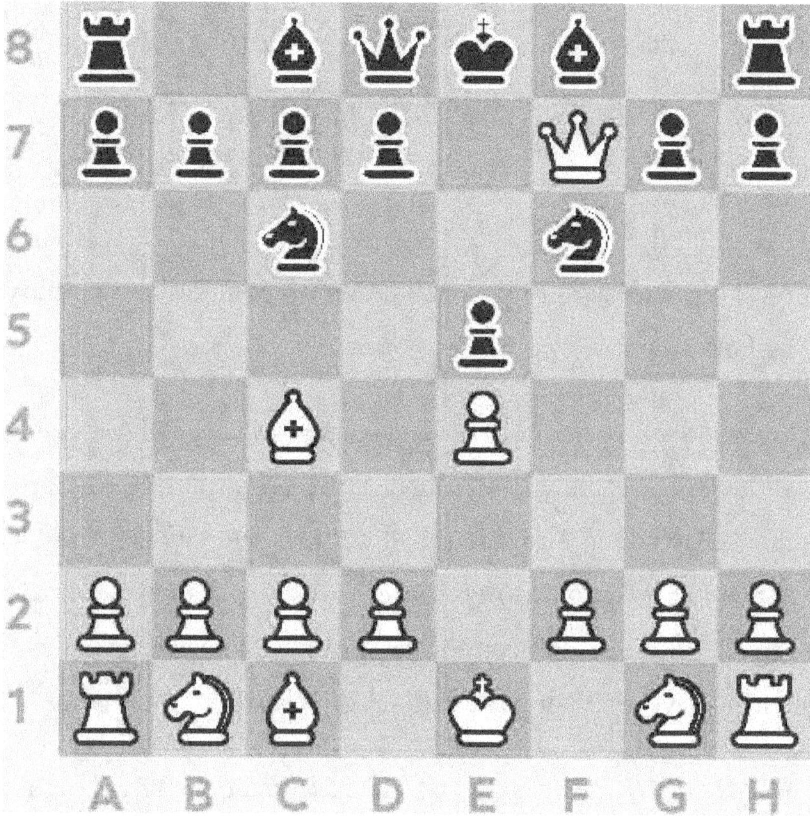

- **Another way to get to the same position**

1. e4 e5 White Pawn to e4, Black Pawn to e5.

2. Bc4 Nc6 White bishop to c4, Black knight to c6.

3. Qh5 Nf6?? White Queen to h5, Black knight to f6 (terrible move).

4. Qxf7# White queen to f7, checkmate.

- **Common mistake**

1. e4 e5 White Pawn to e4, Black Pawn to e5.

2. Qh5 g6?? White Queen to h5, Black Pawn to g6 (terrible move). Black makes a mistake by not protecting the e5 Pawn and opening the a1-to-h8 diagonal.

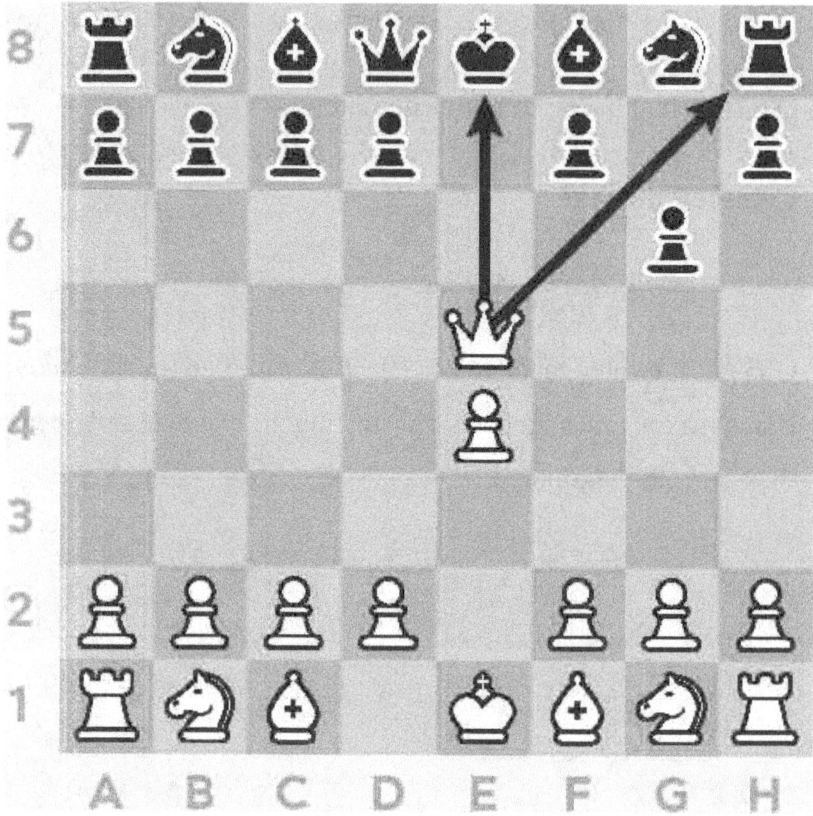

3. Qxe5! Be7 White queen captures Pawn on e5 (good move), bishop
    to e7. White captures the Pawn and attacks the king and rook at
    the same time. Black must protect the king by blocking since the
    king can't run away without being captured.

4. Qxh8 White queen captures rook on h8. White is up 6 points and
    should win the game.

How to Protect against Scholar's Mate

It depends on how White starts the Scholar's Mate: (1) if they bring out the queen first and then the bishop, or (2) if they bring out the bishop first and then the queen.

- **Queen, then Bishop**

1. e4 e5 White Pawn to e4, Black Pawn to e5.

2. Qh5 Nc6 White queen to h5, Black knight to c6. White queen eyes the f7 Pawn and attacks the e5 Pawn. Black develops its knight and protects the Pawn.

3. Bc4 Qe7! White bishop to c4, Black queen to e7 (good move). Bishop develops to c4 and attacks the f7 Pawn, so now there are two attackers (bishop and queen) and one defender (king). Black adds a defender to f7 by moving their queen. It's not a good idea for White to capture the f7 Pawn because they would lose material: 4. Bxf7?? Qxf7; 5. Qxf7 Kxf7 (4. White bishop captures Black Pawn on f7 [bad move], Black queen captures White bishop on f7; 5. White queen captures Black queen on f7, Black king captures White queen on f7), and Black is now ahead by 2 points.

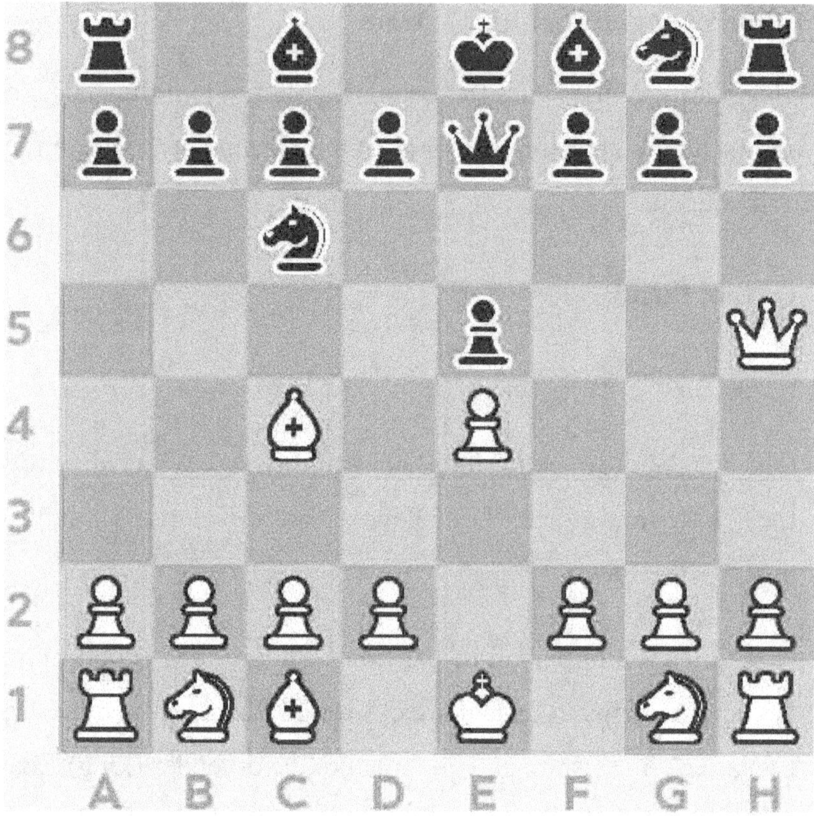

4. Nc3 Nf6 White knight to c3, Black knight to f6. Black has now successfully stopped White's plan and drives off the queen while developing their Knight.

- **Bishop, then Queen**

This one is simple: Black shouldn't allow the White queen to move to h5!

1. e4 e5 White Pawn to e4, Black Pawn to e5.

2. Bc4 Nf6 White bishop to c4, Black knight to f6.

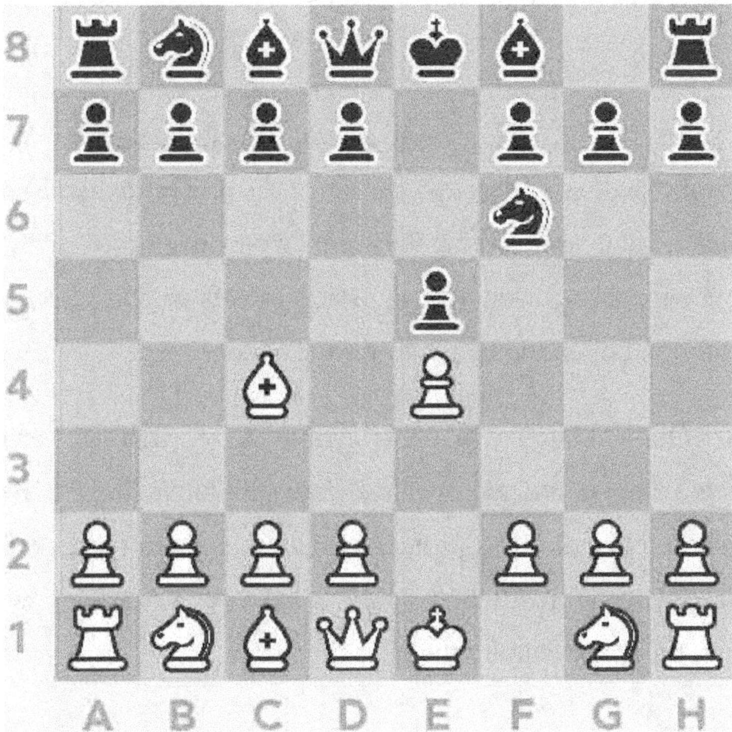

That's all you have to do! There's no reason for White to play the move Qf3 (and not attack any piece at all) or the horrible move Qh5 (which loses the queen).

Fool's Mate

The shortest checkmate is called Fool's Mate, but this one you can't force. Even a beginner would have to be very unlucky to make the worst two moves when there are so many options to choose from in the starting position. Try not to move your f Pawn, whether you are playing White or Black when you're a beginner at chess, or you might accidentally fall for this!

1. f3? e5 White Pawn to f3 (bad move), Black Pawn to e5. White's first move doesn't seem so bad since it's protecting the e4 center square, but it's preventing the knight from developing to f3, and it opens the e1-to-h4 diagonal. Black moves to e5 to protect the center and open the diagonal for the queen.

2. g4?? Qh4# White Pawn to g4 (terrible move), Black queen to h4, checkmate. White makes an enormous blunder! The Pawn on g2 was the only piece to protect against the queen's check. Black checkmates with 2. Qh4#: There is no way to run away with the king, block, or capture the checking piece.

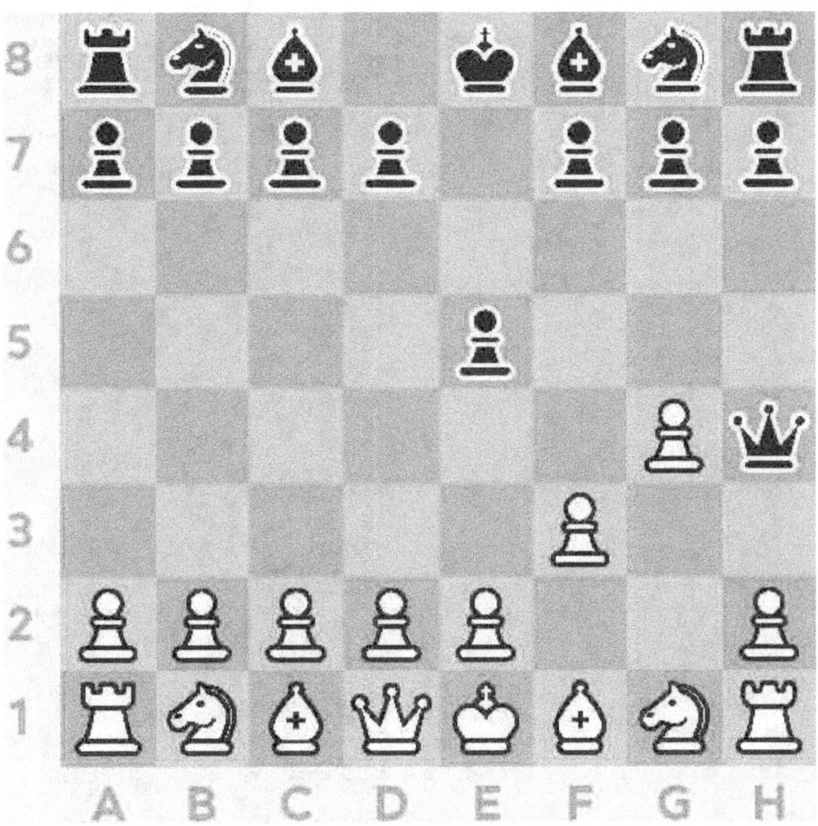

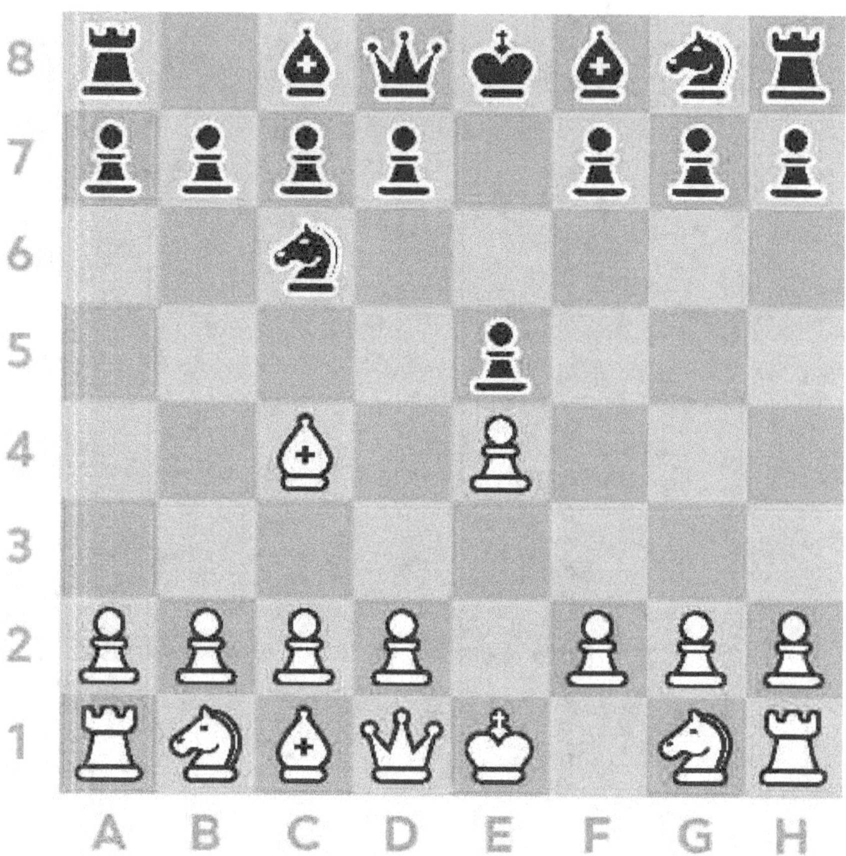

9. Try a Fried Liver Attack

The Fried Liver Attack is a sequence of moves White can use to potentially win material at the beginning of the game. Black has only one correct response to this—the reason many players take the chance of playing it.

1. e4 e5 White Pawn to e4, Black Pawn to e5. Both Pawns control the center.

2. Nf3 Nc6 White knight to f3, Black knight to c6. White attacks the e5 Pawn. Black defends with the knight.

3. Bc4 Nf6 White bishop to c4, Black knight to f6. The bishop develops the position to look at the f7 Pawn. Black develops the knight and attacks the e4 Pawn.

4. Ng5! White knight to g5 (good move). White protects the e5 Pawn and attacks the f7 Pawn simultaneously. White now has two attackers on f7, while Black has one defender. Black's best and only response to defend f7 is 4. d5! Black Pawn to d5 (good move).

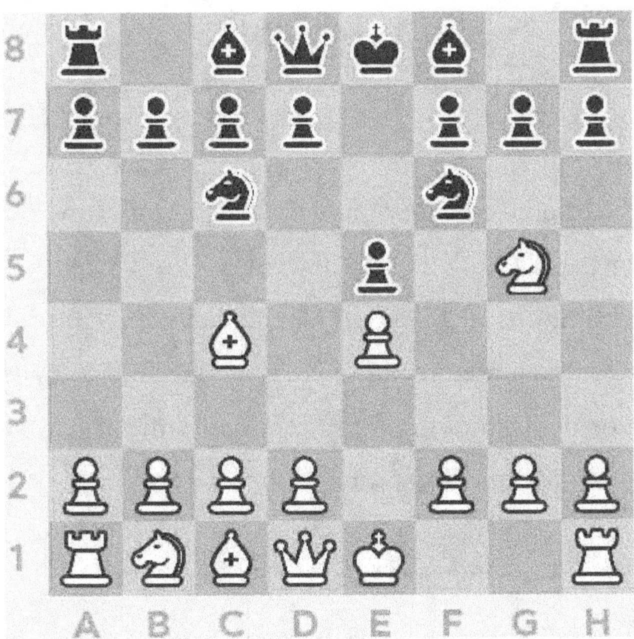

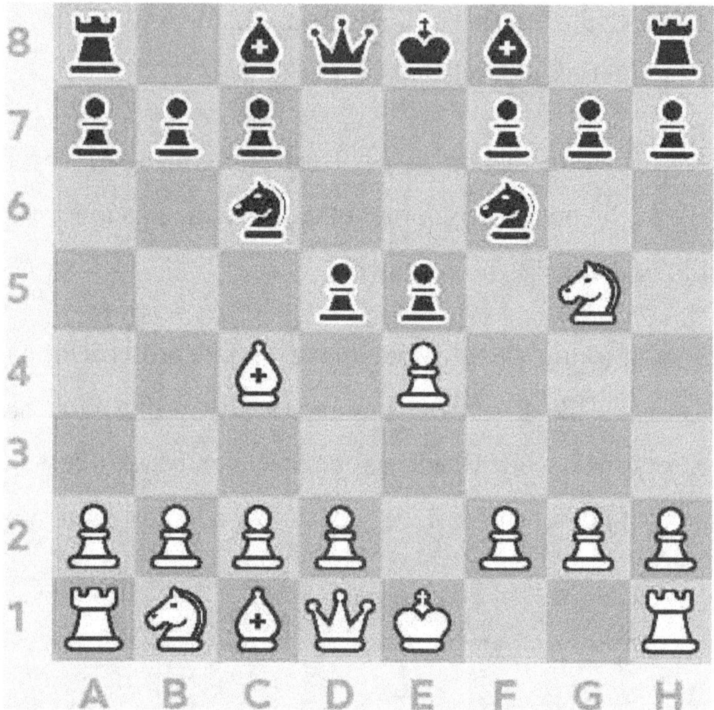

Black blocks the bishop's diagonal, and now there's only one attacker and one defender on the f7 Pawn. The d5 Pawn has two attackers but also two defenders, which means if White decides to capture it, this will end up being a trade. For example, 5. exd5 Nxd5; 6. Bxd5 Qxd5 (White Pawn captures Pawn on d5, Black knight captures Pawn on d5; 6. White bishop captures knight on d5, queen captures bishop on d5).

If Black ignores the Fried Liver Attack

1.   e4 e5 White Pawn to e4, Black Pawn to e5.

2. Nf3 Nc6 White knight to f3, Black knight to c6.

3. Bc4 Nf6 White bishop to c4, Black knight to f6.

4. Ng5! h6? White knight to g5 (good move), Black Pawn to h6 (bad move).

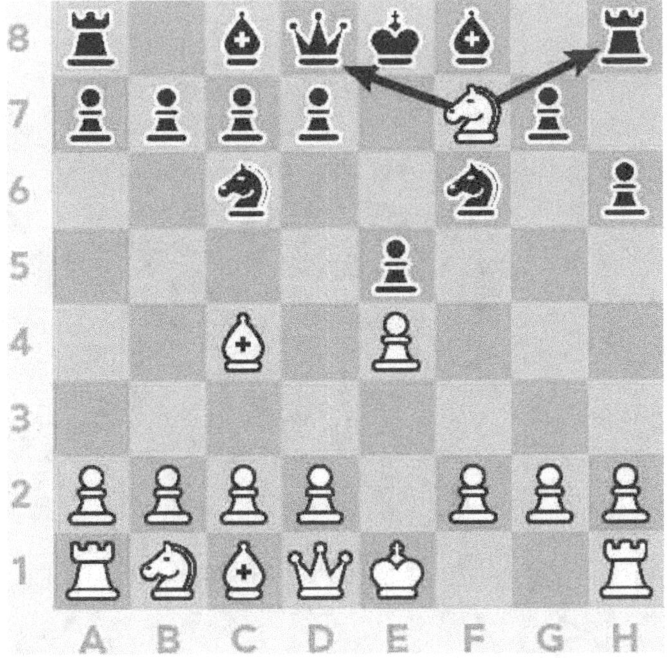

5. Nxf7! White knight captures Pawn on g7 (good move). The Black king can't capture the White knight because it's protected by the bishop; at the same time, the knight is making a double attack (fork) on the queen and rook. Since the queen is valued higher than

the rook, Black will therefore need to move their queen and lose the rook.

6.   Qe7 Black Queen to e7.

White knight captures rook on h8 (Nxh8). White should win this game, being up 6 points from the very beginning!

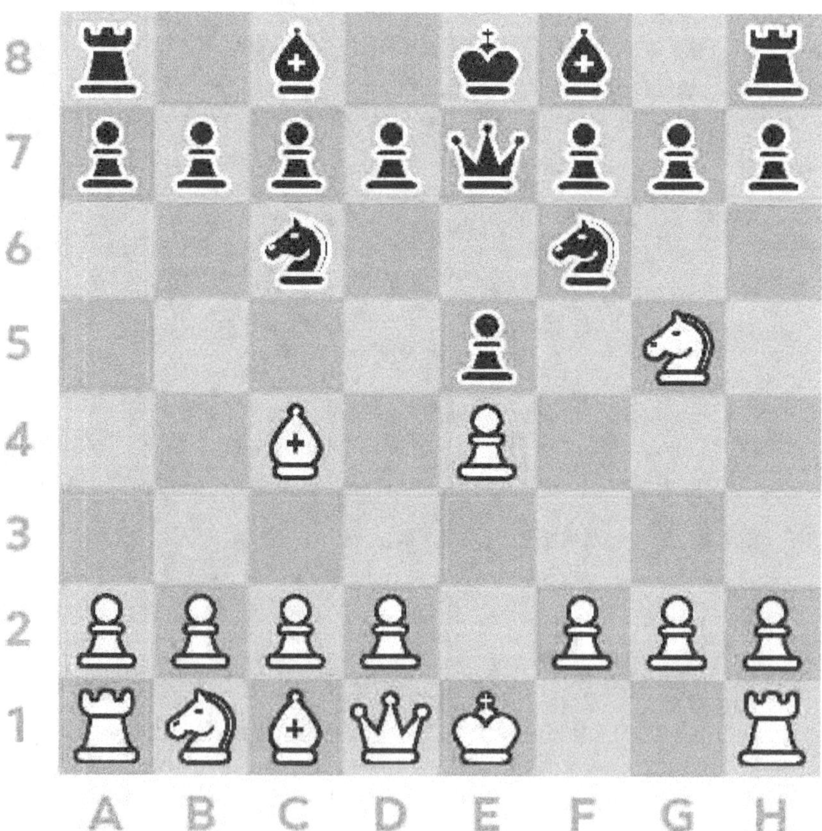

10. Find an Advantage

There are five points I consistently refer to whenever I don't have a tactical move to win material or checkmate. These help me devise a plan whenever one doesn't immediately come to mind:

1. Material Advantage

2. King Safety

3. Pawn Structure

4. Space/Territory

5. Piece Mobility

Material Advantage

This is the most obvious advantage and disadvantage in chess. This is what you should think about if you are on the winning or losing side.

- **Winning Side**

Trade pieces. The more of your opponent's pieces you eliminate, the less you have to worry about.

Trade queens. Get rid of your opponent's strongest attacker, unless you need your own for a winning attack.

Don't be overconfident. "The game isn't over 'til it's over!" Play your best until the end. If you let your guard down, you might make mistakes and change the outcome of the game.

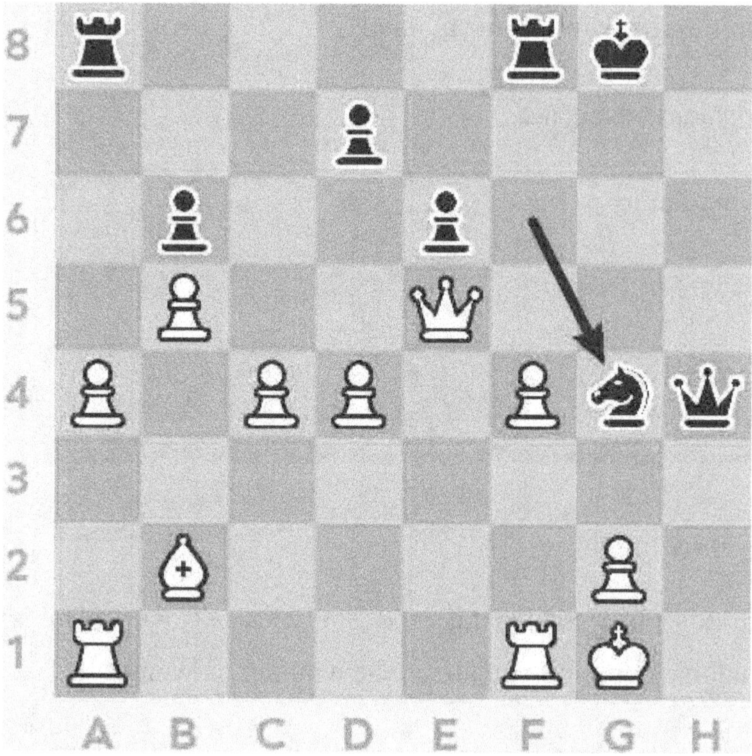

If you're in a bad situation, your king is open, and your opponent seems to have a strong attack, you can always give up some material to get your king to safety.

White is winning by 3 points, but Black's move is a little scary. Black is making a double attack (fork) on the queen and threatening the Qh2 checkmate. How should White defend against this? Trade queens! Black's main attacker is the queen, so since White is ahead in material, White benefits from the trade.

- **Losing Side**

Don't trade. The more pieces you have on the board, the greater your chances to initiate a strong attack.

Don't always look for a way to win. Make sure to always be thinking about draws, too. It's better to tie the game than lose.

Sometimes it's good to sacrifice material to gain access to your opponent's king. You're already losing, so why not make the game interesting with a strong attack? If you're only a Pawn or two down, I wouldn't suggest going overboard, but if you're down a Rook or more, I will go full out!

Below is an interesting scenario where Black was completely winning but didn't see White's threat. But first, let's answer this question:

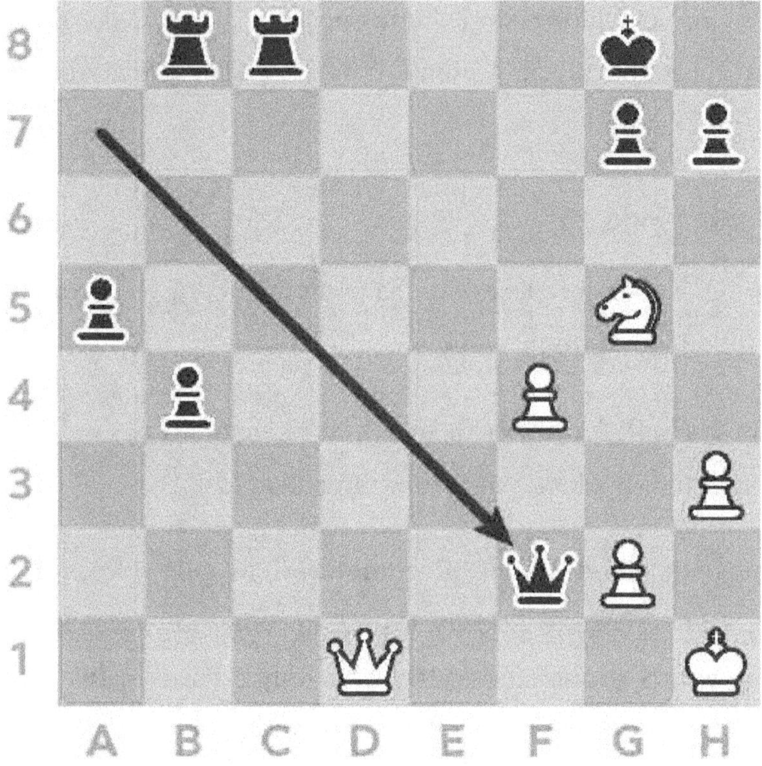

1. QF4?? Black queen to f4 (terrible move). A horrible move that sadly loses the game.

2. Qb3+ Kh8 White queen to b3, Black king to h8. Black can't block the check, or they will just lose material. Black can't play 2. Kf8 because of 3. Qf7#.

3. Nf7+ Kg8 White knight to f7, check, Black king to g8. White brings the knight into the attack! Black's only response is Kg8.

4. Nh6++ Kh8 White knight to h6, double-check, Black king to h8. Same as before, the king can't move: Black king to f8, White queen to f7, checkmate (4. Kf8; 5. Qf7#).

5. Qg8+!! Rxg8 White queen to g8, check (excellent move), rook captures queen on g8. Absolutely amazing move by White. This leads to a smothered mate. Notice: 5. Kxg8 is not possible because the queen is protected by the knight—only the rook can capture it.

6. Nf7# Black rook takes queen on f7, checkmate.

King Safety

Make sure to observe your opponent's king (and yours) at all times! If your king is castled, but your opponent's king isn't, how can you keep theirs in the center and prevent them from castling? Your opponent just moved a Pawn near their king, and now the king is more open, so find ways to destroy the defense and open it even further! Are some of your pieces nearby the king? If so, can you bring in more pieces to start an attack?

Don't forget about your own king! If the king is slightly open, or there are opponent's pieces nearby, this is probably a sign you must go into defense mode!

White is up a Pawn, which is also a passed pawn. A passed pawn is a Pawn that can't be stopped by an opponent's Pawns to get to the promotional rank. The problem for White is that the Black Rook is on a2, which is looking toward the king. We will take a look at this position from both perspectives.

- **Black to move**

Black knows they are down a Pawn, so if they don't do anything aggressive, they will lose the game. The Rook on a2 is very strong and looking at the g2 Pawn, which at the moment is protected until Black adds another attacker. Black notices that most of White's strong pieces are positioned on the queenside, far from the king. If it were Black's turn to move, it wouldn't end well for White!

1. Qg5 Black queen to g5. This adds another attacker on the g2 Pawn and threatens Qxg2#. White can't protect the Pawn with another piece without losing material. (If 2. Qf2 Rxf2, Black will be up 3 points in a Queen-versus-Rook endgame.)

2. g4 Qd2 White Pawn to g4, Black queen to d2. The only move for White. Black continues with their threat on checkmate with their last move: Qg2#. Black queen to g2, checkmate.

There are a few possible endings:

#1

3. Kf1 Qe2+ White king to f1, Black queen to e2, check.

4. Kg1 Qg2# White king to g1, Black queen to g2, checkmate.

#2

3. Qf2 Qxf2+ White queen to f2, Black Queen captures queen on f2, check.

4. Kh1 Qg2# White king to h1, Black queen to g2, checkmate.

#3

3. Rb2 Ra1+ White Rook to b2, Black Rook to a1, check.

4. Rb1 Rxb1# White Rook to b1, Black Rook captures Rook, checkmate.

- **White to move**

In the actual game, it was White to move. Botvinnik found the best move to stop his opponent's threat: 1. Qg5.

1. Qe3! White Queen to e3. The queen protects the g5 square, to which Black is still welcome to move, but when you're down material, you don't want to trade pieces. White also opens up the way for the c Pawn to eventually promote!

Pawn Structure

Pawn structure weaknesses are isolated pawns, double/triple pawns, and pawn islands. There are exceptions when these weaknesses can help in different parts of the game.

- **Isolated Pawns**

Isolated pawns are Pawns that aren't being protected by other Pawns. Isolated pawns are usually an inconvenience because you have to use your strong pieces as defenders when they could be of better use somewhere else. If you want to get rid of an isolated pawn, look for a way to trade it with a Pawn of your opponent's.

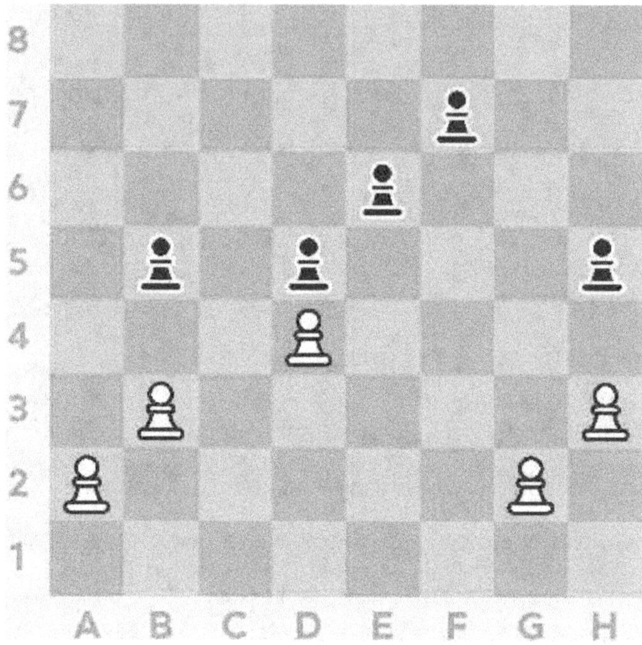

Here, White's Pawn on d4 is isolated. There is no Pawn on the c or e file to protect it.

Black's Pawns on b5 and h5 are isolated for the same reason. Because of this, both sides' Pawn structure isn't the best.

- **Double/Triple Pawns**

Double pawns are two Pawns positioned on the same file. Similarly, triple pawns are three Pawns on the same file. Technically speaking, it is possible to get quadruple pawns in chess, but it is rare. I feel like this is self-explanatory, such as in the following example:

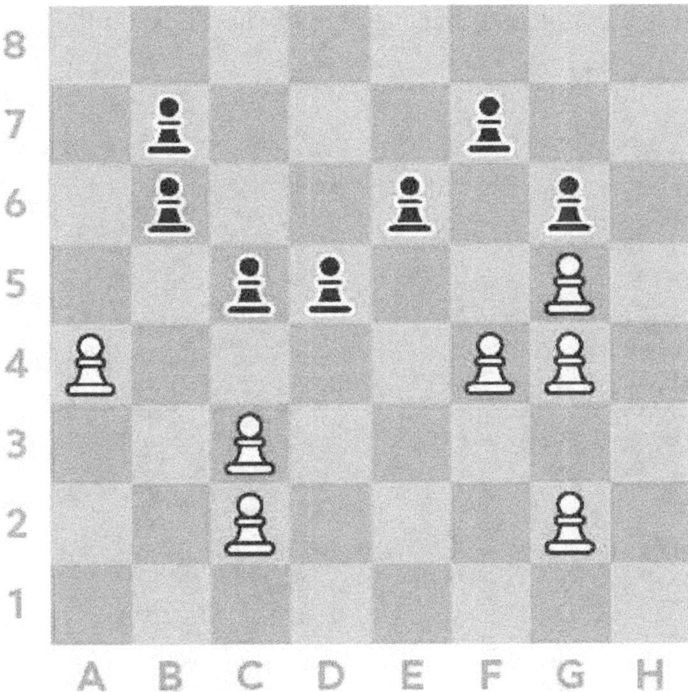

- **Pawn Islands**

While isolated pawns are individual unprotected pawns, pawn islands are groups of Pawns connected by files. The more pawn islands among your pieces, the weaker your pawn structure, as they offer more targets for your opponent. A pawn base is the first Pawn in a pawn chain (or string of Pawns) not protected by another Pawn. The more pawn islands you have, the more pawn bases you have, as well.

- **Two Pawn Islands versus Two Pawn Islands**

White has two pawn islands because there's no Pawn on the e file: (1) a2, b3, c4, d5 and (2) f4, g3, h2.

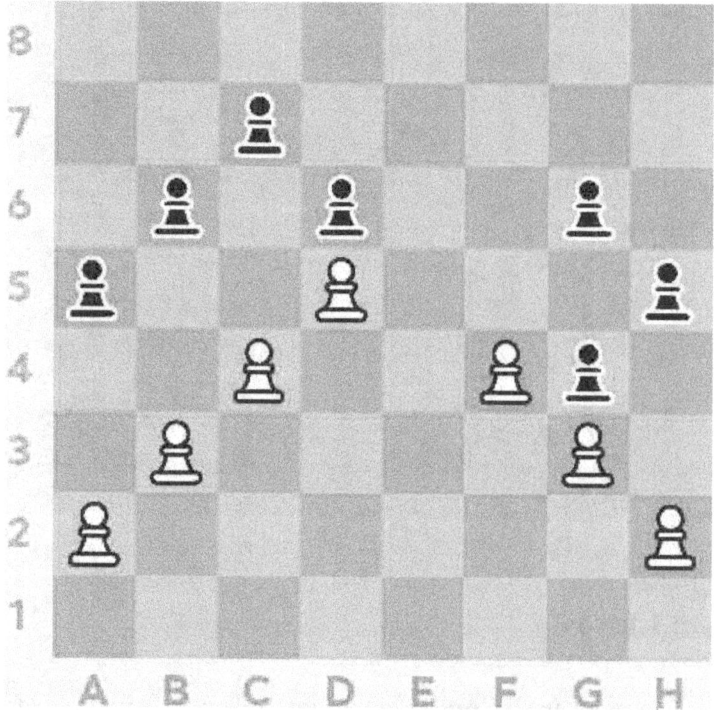

Black has two pawn islands because there's no Pawn on the e file and f file: (1) a5, b6, c7, d6 and (2) g4, g6, h5.

White's pawn bases are a2 and h2. These Pawns are going to be a target, and White will have to use higher-ranked pieces to protect them. Black's pawn bases are c7 and g6.

- **One Pawn Island versus Two Pawn Islands**

Below, White's pawn structure is more promising because it has fewer weaknesses. White has one pawn island and no isolated or double/triple pawns. Black has two pawn islands, which will later become two weaknesses, plus the double pawns are really not favorable, either. One benefit of the double pawns (in this diagram), however, is that the e file is open for a Rook. That's what I mean about every pawn weakness having exceptions. Trading pawn weakness for greater mobility of another piece is worth it most of the time!

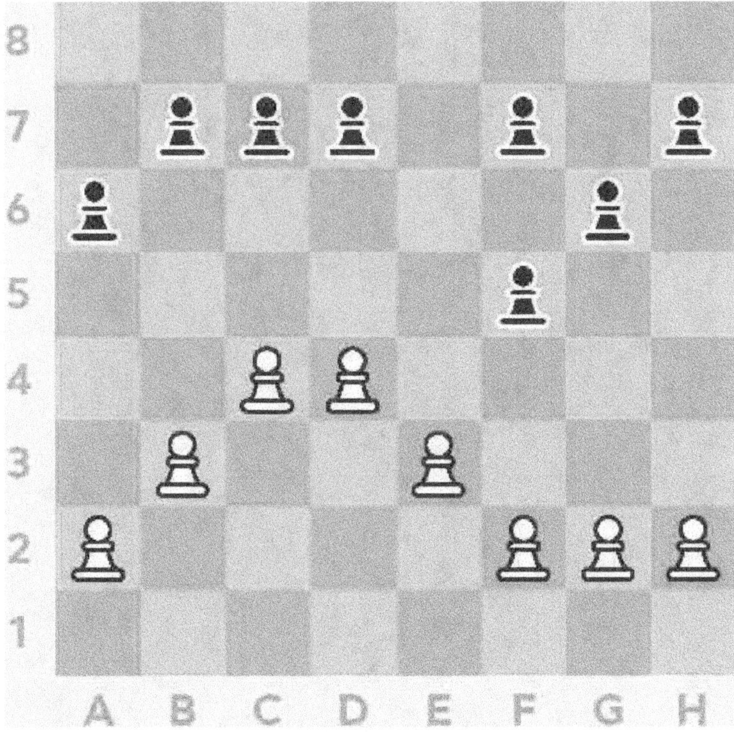

- **Three Pawn Islands versus Four Pawn Islands**

Black's pawn structure is a superior one because it has fewer weaknesses. To be honest, I don't like either side here, but Black's position is slightly stronger. White has four pawn islands, and three of the islands are isolated pawns. The pawn base on e3 is definitely a target, too. Black has three pawn islands, two pawn bases, and only one isolated pawn.

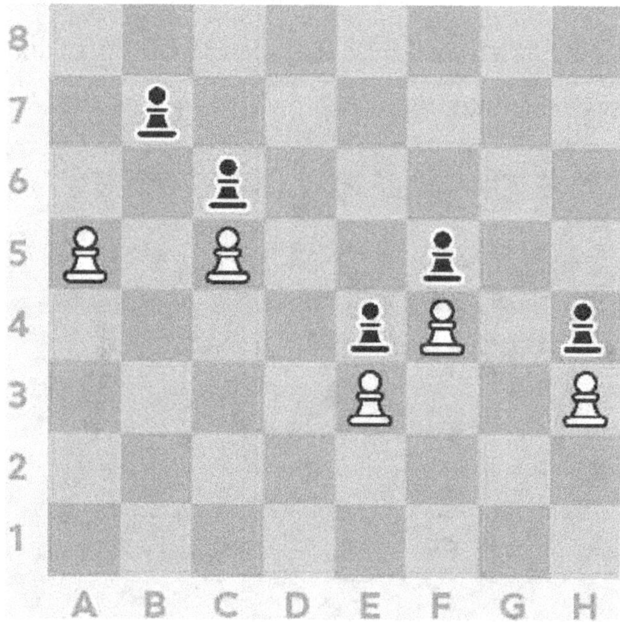

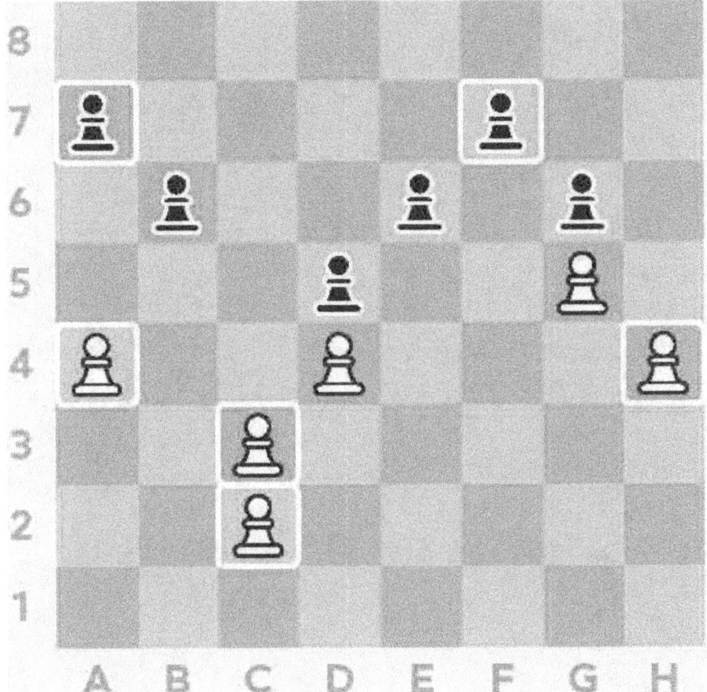

Space/Territory

When it comes to space on the board, the more you have, the greater your advantage in the game.

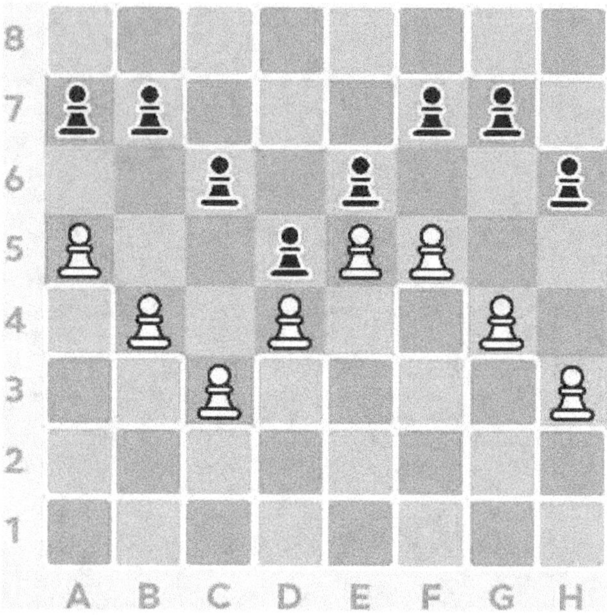

(MORE) SPACE = (MORE) PIECE MOBILITY = (MORE) PLANS

More space on the chessboard equals more movement for your pieces. If your pieces display a lot more movement compared to those of your opponent, this leads to more ambitious plans you can carry out against them. When a side has little space, they can barely move their pieces, and they usually find they don't have many options for moves.

Space can help you distinguish what strategy you are going to choose.

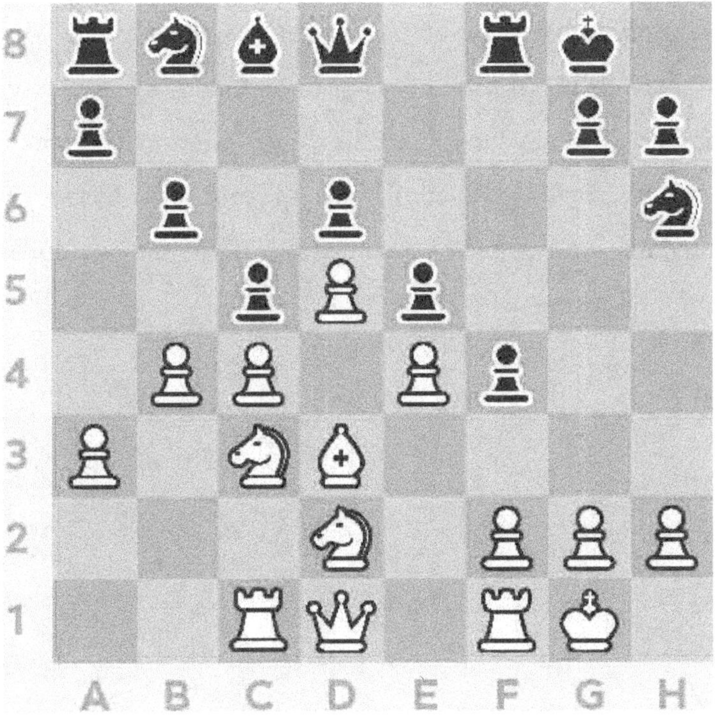

Shown above is a position from one of my games. Playing Black, I have more space on the kingside, so my plan is to form an attack by playing g5–g4. This will weaken my own king but target my opponent's King, as well. White's plan is to play on the queenside since their Pawns are more advanced on that part of the board. In the next section, you will see their position at the end of this game.

Piece Mobility

As stated above, you can now see how space and piece mobility are connected to one another. When analyzing a chess position, it is particularly important to look at all of your opponent's and your own pieces and Pawns. If you notice any of your opponent's pieces have very limited mobility, you may be able to find a way to trap them (take away their only escape routes). What if some of your pieces have limited movement and may be a target for getting trapped? Give them more room! I love trapping Queens—who doesn't?

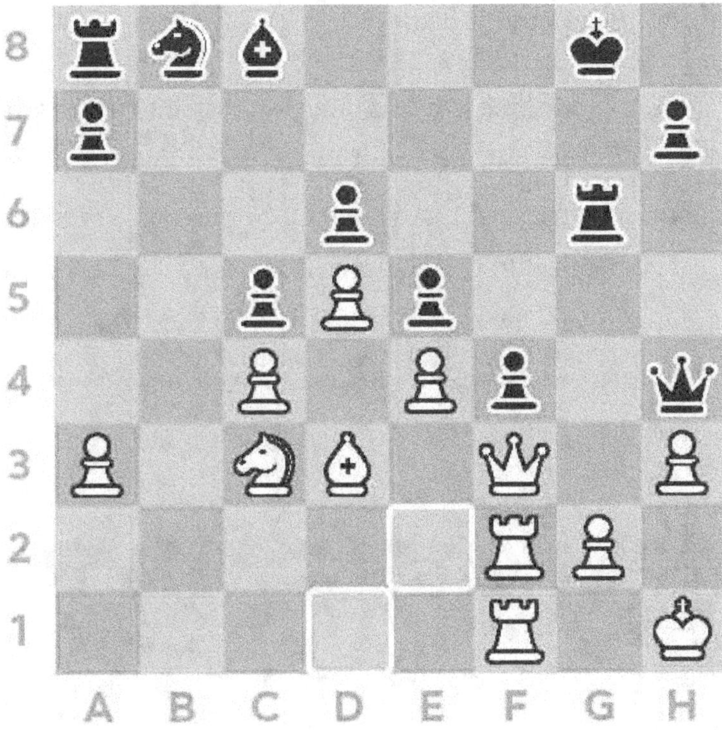

11. Establish and Maintain Tempo

In chess, tempo means reaching the desired position or result in fewer moves than the opponent. So, if a player manages to castle in fewer moves than his opponent, he gains tempo. Initially, white sets the tempo since he makes the first move of the game. But tempo can be gained and lost multiple times throughout the course of a chess game.

12. Develop Pieces by Attacking

Having the initiative is a great way to establish and maintain control during a chess game. For this reason, it is helpful when developing pieces to place them on squares where they directly target the opponent's pieces. Doing so can force the opponent to waste a turn (lose tempo) by moving a piece away from an attack. Doing so causes delays in the opponent's development of pieces making it easier for a player to establish an early positional advantage.

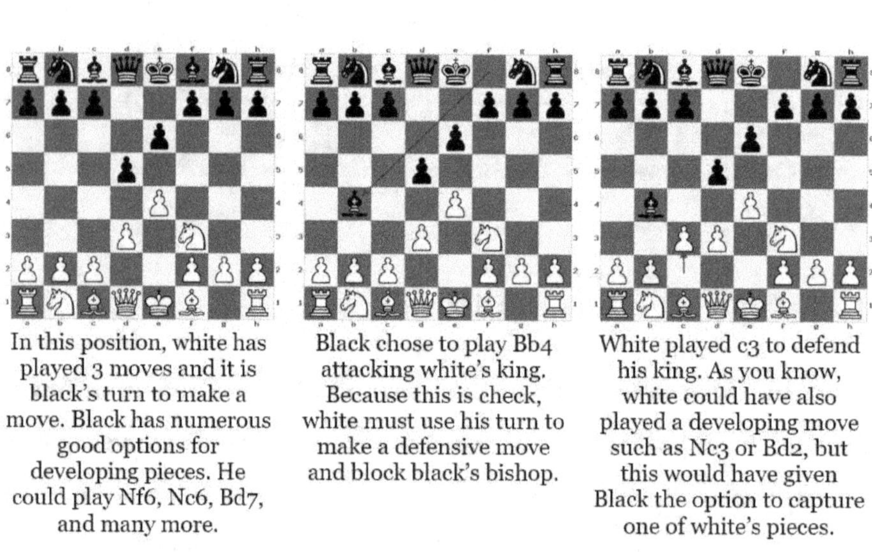

In this position, white has played 3 moves and it is black's turn to make a move. Black has numerous good options for developing pieces. He could play Nf6, Nc6, Bd7, and many more.

Black chose to play Bb4 attacking white's king. Because this is check, white must use his turn to make a defensive move and block black's bishop.

White played c3 to defend his king. As you know, white could have also played a developing move such as Nc3 or Bd2, but this would have given Black the option to capture one of white's pieces.

13. Develop Your Knights First (before Bishops)

As mentioned before, knights are more effective in the center of the board. They help take control of numerous squares and support center pawns.

Consequently, it is advantageous to develop knights towards the center as early as possible. Excellent developing moves for knights are Nf3 and Nc3 for white and Nf6 and Nc6 for black. Generally speaking, it is better to develop knights towards the center and to keep them away from the edges of the board.

Once both knights are developed, bishops can come into the game. Because bishops can move freely along the board's diagonals, they can move across the board more quickly than knights, and thus, it is quicker for them to get to an advantageous position than it is for knights. That's one of the reasons why it's important to develop knights first.

14. Develop with a Purpose

Don't develop just to develop. It is important to place your pieces in the center of the board as soon as possible. However, doing so is a process that requires purpose. Every developing move you make should have a purpose behind it. As you know, there are numerous developing moves you can make, especially early in the opening, but some of these options are better than others. Choosing moves with a purpose helps make sure you don't make a bad move.

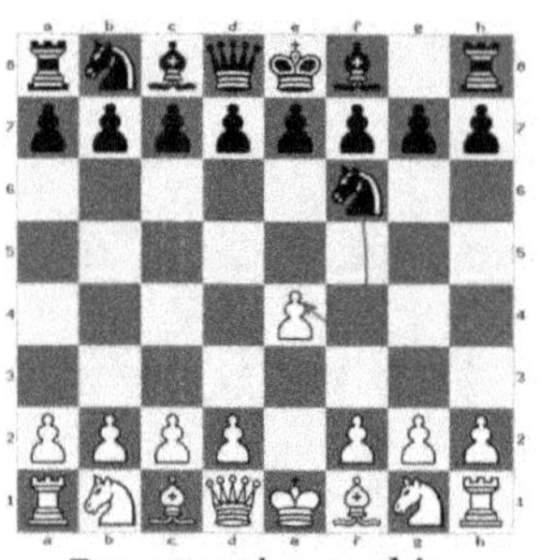

Pay attention to this position. 1 move has been played:
1.   e4 Nf6
It is white's turn to make a move. White has numerous options for developing moves such as Nf3, Nc3, Bc4, etc. However, as you can see, white's e4 pawn is currently being attacked by black's f6 knight.

Therefore, a good way to develop another piece with purpose would be to play Nf3. Not only would white develop one of his knights, he would do so with a purpose: to defend his e4 pawn.

If white chose to play Nf3 instead, he would give black the opportunity to set a material advantage by capturing the pawn on e4.

This is just an example, but most developing moves can and should have a purpose.

15. Castle Early

As mentioned before, castling is the only chess move that allows a player to move two pieces at once. Not only does castling add extra protection to the king by placing it in the corner of the board where it is least vulnerable, but it also develops one of the rooks helping with the preparation of an attack. The sooner you castle, the sooner you can start going after your opponent's king. A good rule regarding castling is to try to do it before the 10th move. Castling kingside is easier since only 2 pieces must be developed to clear the way for the king. It is also considered the better option by many because it puts the king further from the center than when castled on the queenside.

16. Try to Keep Your Opponent from Castling

Considering the benefits of castling mentioned above, it is a good strategy to keep your opponent from redeeming these benefits. By keeping your opponent from castling, you make sure his king remains exposed in the center of the board. It also slows down your opponent's piece development piece. Earlier in this book, we went over the many conditions required for castling. By making sure one of these conditions is violated by your opponent, you can prevent him from castling. For example, by forcing them to move their king before castling, you can make sure they lose castling privileges. Another way to do so is to always keep the squares between their king and their castling rook attacked.

17. Keep the Castle Closed

Once you've castled, it's important that you refrain from moving the pawns right in front of your king. These pawns work as a shield for the king, and moving them might expose it, making it easier for your opponent to attack. So depending on which side you chose to castle on (kingside vs. queenside), keep the pawns on that side in their starting position for as long as possible. You might have to move them at some point to close the way for attacking pieces or to give your king an escape route. But unless you feel it absolutely necessary, don't move them.

In this opening, both players have castled king side. At this point in their game, there is no need for either of them to move the pawns on the f, g, and h files. Until a position later in the game requires them to do so, they should keep their pawns where they are to keep their kings as protected as possible.

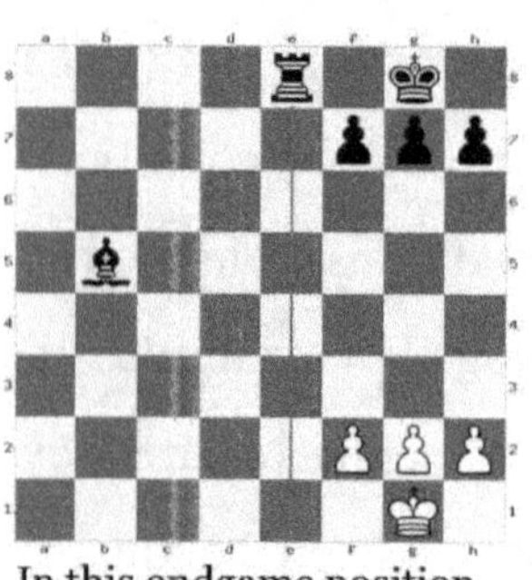

In this endgame position, it is white's turn to move. Black is threatening to checkmate white's king by playing Re1.

So in order to avoid losing the game, white must create an escape route for his king. He does so by playing h3.

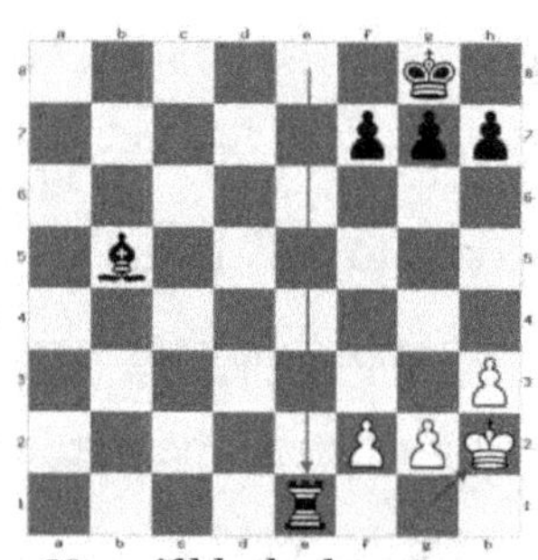

Now, if black plays Re1, white's king can escape to h1 and keep the game going. This is the kind of situation that require the pawns in front of the king to move.

18. Connect Your Rooks

As mentioned before, rooks are "connected" when all the space separating them is clear. In other words, when all the pieces between a8 and h8 have been developed, rooks are connected. Generally speaking, two pieces are connected when one has the moving power to defend the other.

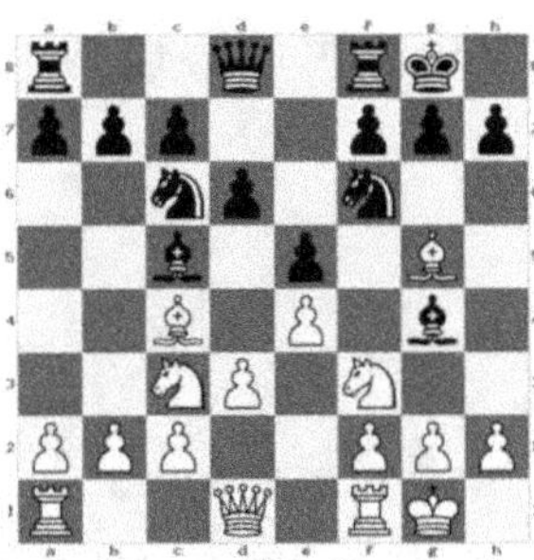

In this position, white has castled king side and is one step away from connecting his rooks. To do so, he must develop his queen.

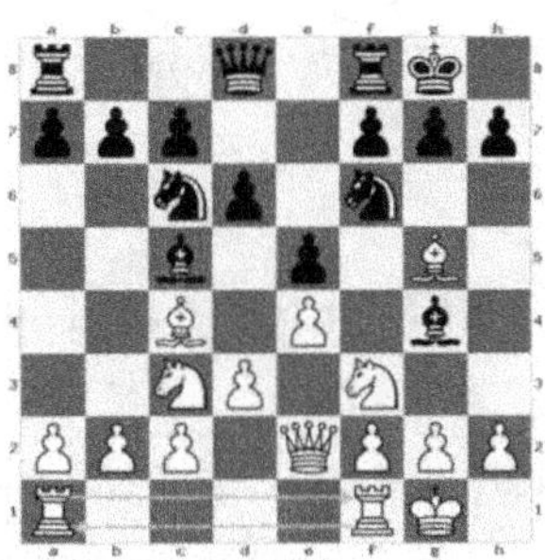

White played Qe2 His rooks are now connected, marking the end of the opening phase of the game.

Connecting pieces is usually a good practice for attacking an opponent. It allows a player to elaborate an attack and put pieces in place for it while reducing the risk of losing material.

It is black's turn to make a move. Black's bishop on f3 is under attack from white's f1 rook. Pay attention to black's d5 pawn. It is currently cutting the communication between black's queen on b7 and his bishop on f3. So, black should play d4 to reconnect these two pieces.

After black plays e3, the connection between black's queen and his bishop is established, and it is no longer a good idea for white to capture black's Bishop on f3 as it is defended by black's queen.

If white plays Rxf3, he will surely lose his rook as it becomes the target of black's queen.

So, when possible, a 4th way to defend a piece under attack is to connect it to another piece on the board.

19. Don't Bring the Queen out Too Early

It is recommended to not develop the queen before other less valuable pieces. The reason is that the queen is the most powerful piece on the board; its survival is very important for one to win the game. Bringing it into action too early might expose it to capture the opponent's less valuable pieces. You would then have to waste an important number of defensive moves in order to protect the queen, which would make you late in the development pieces (give the opponent a development advantage). This would give them a significant advantage and drastically increase their chances of victory.

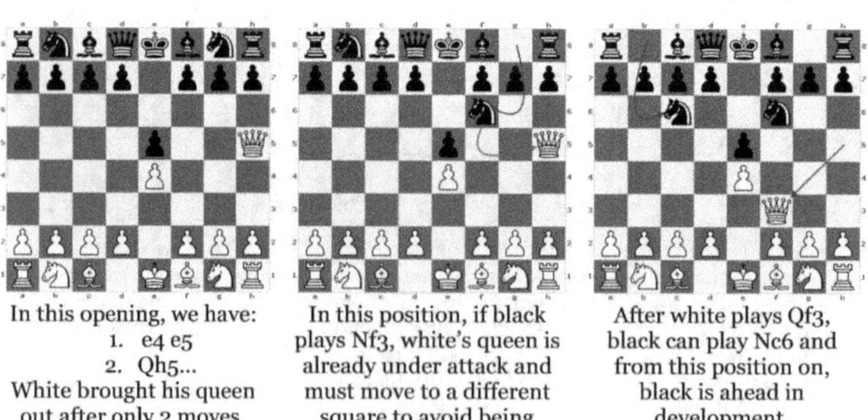

In this opening, we have:	In this position, if black plays Nf3, white's queen is already under attack and must move to a different square to avoid being captured.	After white plays Qf3, black can play Nc6 and from this position on, black is ahead in development.

In this opening, we have:
1. e4 e5
2. Qh5...
White brought his queen out after only 2 moves.

The same applies to your rooks. As the second most powerful piece on the board, you should wait until having developed other pieces before bringing your rook into the game.

20. Try Not to Move Pieces Twice In a Row

As you know by now, it is very important to quickly develop pieces while maintaining a good tempo. A good way to achieve both results is to avoid moving the same piece twice during the opening. In other words, after developing a piece, a player should focus on making other developing moves and consider moving already developed pieces later in the game. Some cases might require moving the same piece twice, although this is somewhat rare.

In this position, it is white's turn to make a move. White developed his king side knight in the previous turn by playing Ne2. Black responded with Nc6.

Moving the same knight again by playing Ng3 for example, would not be recommended as it would slow down white's development and affect his tempo. If black played Nf6, he would be ahead in development.

A good alternative move would be Nc3 developing a new piece and maintaining tempo.

There are many more general tips and strategies for openings to help you improve your chess game, but the 20 mentioned above are some of the most important ones.

Chapter 3. Tactics to Support Your Strategies

Chess is more about tactics. Therefore, you have to improve your chess strategy. "Chess is ninety-nine percent strategic," said Richard Teichmann, a celebrated German chess instructor, in 1908, and he was correct.

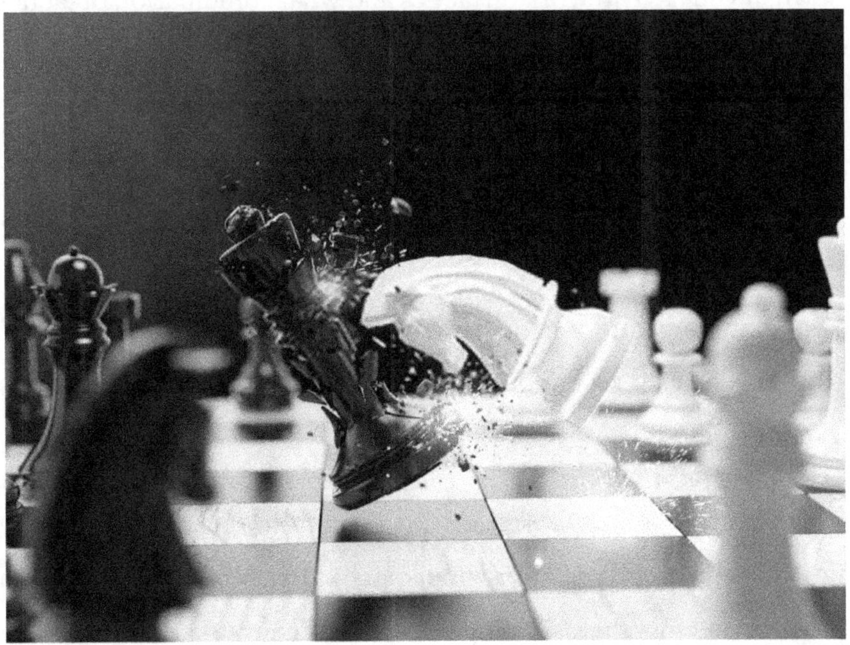

Tactics play an important role in all chess games and are the essential elements of any strategy. A strategy is a general game plan, sentences about where you need to be at any point during the match. The tactic is used to implement a strategy. Your opponent will not sit and watch you lead your strategy without objections. The tactic is used to force your opponent to accept your moves. The tactic shows one or more moves to gain a short-term advantage. These are fundamental steps in advancing a general strategy.

Many tactics can be memorized or learned and applied if necessary. By setting more tactics, the player's arsenal becomes stronger. Learning new tactics, when you have to work on tactics and what combination of tactics works best together is a continuous activity throughout a player's life.

Here are some common tactics that are worth practicing:

Tactic 1: Battery Attack

When you think of Battery Attack, imagine an ordinary electric battery; each battery is an electric battery. If you want more control, you need to gather more cells.

In chess, a battery attack is formed on the rows (ranks and files) by accumulating rooks and queens, while, diagonally, a bishop and a queen are stacked or put together to give power to the enemy. The next day.

Both armies can show each other batteries that.

However, suppose the White battery consists of 2 rooks, while the Black battery consists of 2 rooks and a queen. Black has a more robust battery, and White is wise to consolidate his 2 rooks or to prevent an attack. Black battery at this time.

Tactic 2: Block

The block is a defensive tactic and is used whenever a bishop, a rook, or an opposing queen manages to control their king.

What you can do is place a pedestrian or piece of land in the middle of the rival attacking unit and the king- block the check and protect the king at least temporarily.

Be aware, and even if your lock unit is provided by the King or other infantry or support units, the attacker may decide to surprise the blocker in a sacrificial maneuver that has been recorded to remove some of your king's defense.

Tactic 3: Authorization

It is also known as "clean cleaning," which better describes what's going on here.

Imagine you want to place a piece on a particular square; this reinforces the attack you ride. The issue here is that one of your pieces is already in that square.

The issue here is that moving that piece leads to its recording. However, because of the superior position you can get by getting the other piece in that square, it is worth "clearing" that block of the piece, accepting its sacrifice, to compensate for the damage you want. Bring (improved position).

Tactic 4: Angry

Decoy involves sending a rifle or sniper rifle to a particular area as a sacrifice to capture the enemy.

After that, your "real" goal is reached, as the "wait" piece gives you the chance to grab the enemy king or exploit another major part of your rival (usually the queen) to take hold.

Tactic 5: Deviation

Imagine throwing a stone with enough force behind it on another rook. When it strikes, a stone with less power "deviates" from its resting position. In chess, you can create an attack, and there is enough weight behind it (as supporting pieces) to attack the enemy's position at a certain point, and this forces your actual target, just as the king's enemy, to be forced. It escapes from your current. The position, the "lost" king moves away and puts you in a stronger position.

The famous continuous Checkmate is known as "Legall's Mate," which contains Deflection. In move 6, it happens when White's bishop turns the f7-black gun, which will definitely place the black king back in control. To protect the White knight from the "e5" field, the Black king has no choice but to move—this part deviates from the "e7" field (and Checkmate follows White's next move).

Tactic 6: Discovery of the Attack

This tactic requires the cooperation of the two pieces. One will be in front of the other; one of the backs of the hidden unit is waiting to be discovered or "discovered." At the selected moment, the piece moves forward—basically to launch an assault to another pawn/enemy piece, and the remaining track is revealed, attacking another pawn or enemy part (this is not the king's enemy) more on this, in the next tactic.

Following the attack, the opponent will have to choose to save or attack the pedestrian; to save, another attack will attack you. On a defensive note, before making the next move, read the painting and look at your opposition's pieces. If you see two pieces in the neighboring squares (as well as each sitting diagonally, side by side), first look back one by one and follow the down line of the army on the path of your army. If you hit one of your pieces and if the enemy's opponent is within range of your other pieces, your opponent may try to destroy you with a full-blown attack.

Tactic 7: Discovery Revealed

It includes principles similar to the attack detection standard. The main contrast is that the adversary of the ruler is one of the pieces assaulted. Since the king is assaulted, he will be "analyzed," this implies the king must be ensured no matter what.

The checks were intended to catch the "other" adversary casualty.

Tactic 8: Fork Assault

At the point when a pawn or piece assaults (at least two) units with a solitary move, it is known as a "fork assault."

Fork assaults can be "relative" or "outright."

The relative forks assault at least two enemy units, but not the adversary ruler.

Outright forks assault at least two adversary units, and this time, one of the foes IS's pieces is the ruler. At the point when this is a relative assault, the player can pick which gathering to spare and which to leave helpless against the aggressor.

At the point when this is an outright assault, the player's top dog must be ensured, as in check. The pawn/pieces/different pieces are assaulted by their destiny.

Tactic 9: Intermezzo/Zwischenzug

In English, both "Intermezzo" and "Zwischenzug" mean "moderate movement."

This tactic requires a bit of deception.

First, you make some wrong move, and your opponent reacts to it—this is the "middle move," then you make a move.

Tactic 10: Pin Attack

The pins include attacking a piece of pedestrian or less valuable furniture in the face of a more valuable piece behind it.

If the indispensable piece is the enemy king, it is known as the "absolute needle," and the pawn, the most valuable piece that is in front of it, you cannot move to escape the threat - you can act against it. King. If it happens that an indispensable piece is just another piece, the less valuable piece may go out of its way, but it is often still in place to prevent obtaining a more valuable piece.

Tactic 11: Sacrifices

Sacrifices are intentional attempts to play on foot or piece by piece in such a situation that will be occupied. But instead, it is to compensate for your damage or to capture one of the enemy's pieces or ability to win, developing leadership.

Tactic 12: Brooch Attack

The skewer is the opposite of the pine.

When a brochure takes place, the attack is on a more valuable piece, which is a pity that you can stand in the face of a Pawn or more valuable piece.

The idea behind skewer is to set aside the indispensable piece so you can get any victim it stands for.

Like Pin, the attacks from the brochure could be "more or less" or "absolute."

Tactic 13: Trap

Setting the trap can be a gamble.

If your opponent identifies your hidden program, you will probably lose a move.

Windmill Attacks gets its title from an observer, who watched the game in a game and likened it to the rotating blades of a windmill, and the name has been used forever in these rare tactics.

What happens is that, due to the placement of a spare part, the attacking piece puts the king's enemy in control, who must move to rotate to exit.

It allows our attack unit to capture one of the enemies, this allows the king's enemy to retreat, but the attacker re-examines the king, who is forced to take a turn. Pass to transfer to security, but this allows the attacking unit to capture another enemy.

This series of rotations of surveys and recordings is the reason for this sequence, which is called an air mill.

Tactic 15: X-Ray Attack

When we refer to X-rays in Chess, we resemble Superman's X-ray vision. The term "X-ray" was used to describe a piece's ability to "see" through objects with the goal that they could focus their opinions on whatever happened in it.

It is more of a "threat" than anything else because you cannot target the X-rays. After all, the object (pedestrian or another object) blocks the path.

In total, there are three different forms of X-Ray Attack. Two of them are full of "offensive" attacks, while the third includes "defensive" and "offensive" elements.

- Type 1 is another term for skewer attack.

- Type 2 threatens with one or more significant pieces.

- Type 3 defends intimate parts by one or more enemy pieces.

Mistakes to Avoid

Carelessness can bring undesirable results. This is the same with chess. Some players don't lose because they lack the skill. Rather, they are not very careful with their moves and end up doing something disadvantageous for them. This can eventually cost them the game.

Here are the three common accidents in chess that every chess player should be aware of. Knowing these will allow them to spot possible scenarios that will lead to it and eventually avoid it.

Losing Your Piece for Nothing

Capturing one of your opponent's pieces (or having one of yours captured) is a normal part of the game. It is common for players to bait their opponent to capture one of their pieces because another piece is guarding that square in case it is captured. If you should lose a piece, the opponent should also lose one of theirs. However, it can be costly for a player to lose any piece but don't get anything in return. This situation can be treated as if they gave something to the opponent for free.

Here is an example:

In the image above, it can be seen that White moved his knight to f7 and threaten Black's queen and rook. However, what White fails to notice is that the square where it landed is guarded by Black's king. Since the knight is not guarded by any other piece in case of capture, Black can capture the knight for free!

Sometimes, hanging pieces are not that obvious. The image below appears like a normal opening sequence for the game.

However, if one is to look closely, White can capture Black's rook on the other corner of the board. Free 5 points for White! Even if Black decides to block the bishop, it will surely be able to capture another piece before getting captured.

A player should look at all areas carefully so that having a "hanging" piece can be avoided. Obviously, this is a mistake that you want your enemies to commit, as it can give you an advantage.

Losing a Piece with Higher Value

It was mentioned earlier that having one of your pieces captured for free is bad. This opens the idea that you should at least get something in exchange for that piece. However, this doesn't mean that we should settle for pieces that have a lower value. Capturing an opponent's piece with a lower value than ours is not good either.

If the game drags on and this trend of getting an "uneven trade" continues, the player whose higher value pieces are traded for lower value ones will end up with a deficit and eventually lose the game because their power is significantly reduced.

The image above shows that White took out his queen way too early. Black responded by moving his pawn to d5 and opening the bishop for capture. If White decides to take on the bait, his queen will be easily captured even if Black's bishop has been sacrificed. By simply referring back to the point value of pieces, it can be seen that White will lose 9 points while Black will only lose 3.

Players should remember that even though capturing multiple pieces can help them get an advantage, being able to target and capture high-value pieces is still better. Trades should be based on the value of the piece, not on how many can be captured in exchange for a high-value piece. Even if you are able to capture three pawns, it would still be an uneven trade if your queen or rook gets captured because of those small pieces.

Your Pieces Are Outnumbered

It is common for a player to offer one of his pieces as bait so that he can capture anything that attempts to capture that piece, especially if a piece with a higher value will be involved. However, there are times when a piece is not guarded enough. If the opponent sees an area that does not have enough protection, he will surely exploit that weakness. He may even be willing to sacrifice some of his lower value pieces just so he can capture your higher value pieces. In this case, it's important to defend possible weak areas more. By doing this, your opponent will hesitate in capturing one of your pieces.

The image above is an example of the four-move tactic that can win games for White. This tactic exploits the weakness of the pawn in f7. With White to move, it can be seen that the said pawn is only protected by Black's king. However, that same piece is threatened by White's bishop and queen. Since the defense for that area is not sufficient, White can capture that area without any problems, get protected by the bishop, and checkmate the opponent.

Conclusion

You have to note that this is just an introductory book on chess. Every point we have presented to you up to this point in this book are those we deem important to you as a beginner. There is no need to fill your head with more information that may end up leaving you more confused than you were before you picked up this book to read.

So, what should you do with the little information you have gotten so far? You should put the information into practice. Get yourself a chessboard and get familiar with all the components of the board. Try to memorize all the squares, ranks, and files. Once you are done doing that, the next thing you should do is to arrange the chess pieces on the board correctly.

Start by playing with yourself—you will need to play the White and Black pieces yourself. This is a great way to perfect the new skill you just learned. The more you practice with an actual chessboard, the more you will master how the different pieces move on the board. While playing, make sure you apply all the techniques you have learned in this book. Make use of forks, pins, and the many other techniques we have talked about.

If you cannot get your own physical chessboard immediately, another good way to practice what you have learned is to download a chess game on your mobile phone. There are many free mobile chess games you can download and start playing immediately. Just visit the app store for your mobile operating system, type "chess" into the search bar, and download one of the games that will pop up in the search results. Make sure you read reviews before choosing the one to download.

Some online games will allow you to pair with another player online. Once you have paired with someone, you can compete with this person and win virtual points. Participating in such competitions is a great way to sharpen your skills.

However, when you are just starting, I would recommend that you shun online competitions entirely. You need to learn on your own first before engaging in competitions. Many mobile chess games allow you to play with the system as your opponent, and that's one of the easiest ways to learn how to play chess.

Open the game you have downloaded, select the option that allows you to play with the system, choose a side, either White or Black, and then start playing. When doing so, you should not only pay attention to what you are playing, but you should also pay attention to the moves that the system is making against you. You will learn a lot from the system's moves. Keep practicing and improving your skills, and don't forget that chess can be quite addictive.

Remember, this book has given you the basic foundation you need; it is left for you to leverage this foundation and turn yourself into a chess master. Chess is like music; no one would teach you how to play a musical instrument such as the guitar from beginner to advanced level. No, most instructors would show you the basics—the music progressions—and it is then left for you to figure out the other things that will make you a great guitarist.

The same applies to chess; no one will teach you the hundreds of opening moves and techniques out there. However, someone can furnish you with the fundamentals, just as we have done. It is then left for you to take the basics given to you and flesh it out.

That being said, there are tons of materials you can find about some of the moves, techniques, tactics, defense methods, etc., which we have introduced in this book. For instance, chess opening moves already have thousands of texts written on that single subject. You will also find full books that are only dedicated to chess tactics—pins, forks, etc.

What all this means is that you cannot get all the knowledge at once—you will have to explore and learn more on your own. As you play more, you will grow inquisitive, and this is what will make you try to seek out more information on how to surmount a particular challenge you face.

Lastly, whenever you are playing, don't feel shy to refer back to any section of this book for clarifications on the best way to make a move. This book is like a reference guide for beginners—so you should be able to refer to it from time to time.

Glossary

Absolute pin. When a piece is pinned to the king and cannot move because the king would be placed in check.

Alekhine's defense. A chess opening with a position typically occurring after 1.e4 nf6 2.e5 nd5 3.c4 nb6 4.d4 d6.

Asymmetrical pawn structure. A pawn structure where the pawns are not symmetrical where one player has a pawn on a file where the opponent does not.

Attacker. A piece that attacks an opposing piece.

Back rank. Another name for the first rank, the rank closest to the player behind the pawns in the game's starting position.

Back rank mate. It's a checkmate that occurs when the king is attacked on the back rank by a queen or rook, and his escape is blocked by his own pawns.

Backward pawn. A typically weak base pawn on a half-open file that may be easily attacked by the opponent's pieces, especially the rooks.

Bad bishop. A weak bishop usually blocked by his own pawns that is a permanent disadvantage.

Base pawn. The pawn in a pawn chain that is closest to the player and has no pawn to protect it.

Bishop pair. The advantage of having both bishops typically against one bishop and one knight or two knights; worth about half a pawn.

Blitz. A rapidly played game where each player typically gets five minutes to complete all his moves or lose on time.

Blumenfeld's rule. Writing the move down on a score sheet before playing it so that you can double-check the move first in an attempt to avoid making an obvious blunder.

Candidate moves. Reasonable moves that make sense without calculation, a player's main choices for his move.

Capture. When a piece moves to a square where an opposing piece is resting and removes the opposing piece from the board.

Castling. A special move that usually occurs in the opening, getting the king out of the center and developing the rook, and the only time a player can move two pieces in one turn.

Center. The middle of the board that includes the squares e4, e5, d4, and d5.

Centralizing the king. Strategically moving the king toward the center of the board where he can join the battle, typically occurring in the endgame.

Closed pawn structures. A position with pawns blocking the movement of the pieces.

Combination. A series of moves that are played in an exact sequence to gain an advantage.

Counterattack. An attack mounted by the player who is defending.

Cutting off the king. A barrier created by a rook or queen in a file or across a rank where the opponent's king cannot move across because he would be moving into check.

Dark-squared bishop. A bishop that moves on the dark-colored squares.

Defender. A piece that defends one of its own pieces.

Deflection. A tactic where the defender is chased away from protecting a critical square.

Developing a piece (also called development). When a piece (knight, bishop, rook, or queen) moves off its starting square to a better square, increasing its power.

Discovered attack. Moving a piece and attacking an opponent's piece with a bishop, rook, or queen hiding behind the piece that was moved.

Discovered check. Moving a piece and checking the opponent's king with a bishop, rook, or queen hiding behind the piece that was moved.

Double attack. A discovered attack where the moving piece also attacks an opponent's piece.

Double-check. A discovered check where the moving piece also checks the king.

Doubled-isolated pawns. Two pawns lined up vertically on the same file with no pawn on a file next to them that can protect them.

Doubled pawns. Two pawns lined up vertically on the same file that has a pawn on a file next to them that can provide protection.

Draw. A chess game ending in a draw.

Endgame. The part of the game when only a few pieces are still on the board.

En passant ("in passing" in French). A special pawn capture that exists for only one move and occurs when a pawn on the fifth rank captures an opposing pawn on an adjacent file that advanced two squares forward as if it only moved one square forward.

En prise ("in take" in French). Refers to the piece that can be captured.

Exchange. An equal trade of pieces in terms of material.

Exchange sacrifice. To voluntarily trade a rook for a bishop or a knight, also known as giving up the exchange.

Gaining a tempo. Gaining a move.

Grabbing pawns. A risky way to win pawns, since the opponent may gain piece activity.

Fianchetto. Developing a bishop to the square directly in front of where a knight starts the game where the bishop is on the longest possible diagonal on the board and can attack two of the center squares of its color.

Files. Columns on the chessboard identified by a letter from A to H.

Forcing moves. A series of moves (usually checks) that force a response from the opponent, keeping him from carrying out his plan.

Fork. A common tactic that occurs when a piece attacks two or more pieces at once.

Half-open file. A file where one player has a pawn and the other player doesn't have a pawn.

Hole. A weak square that cannot be attacked by a pawn, ideal for an opponent to occupy with a piece.

Initiative. Having the attack and being able to dictate the direction of the game.

Insufficient mating material. When a player does not have enough material left on the board to force a checkmate.

Isolated-passed pawn. A pawn that has the weakness of being isolated but the strength of being passed.

Isolated pawn. A weak pawn that has no pawn on a file next to it.

Interpose. To block or put in between.

Kingside. The half of the board where the kings begin the game (the files e, f, g, and h).

Kingside majority. Having more pawns on the kingside than the opponent.

King's Indian defense. A chess opening with the position typically occurring after 1.d4 nf6 2.c4 g6 3.nc3 bg7 4.e4 d6.

Lever. The square where a pawn can capture the opponent's pawn to open up the position.

Light-squared bishop. A bishop that moves on the light-colored squares.

Losing a tempo. Losing a move.

Main line. A common sequence of opening moves where both players are playing correctly and the position is fairly even.

Major pieces. Rooks and queens.

Material. The sum of the values of the pieces.

Mating material. Having enough material to force checkmate.

Mating net. When the king's flight squares are eliminated, making checkmate possible.

Middlegame. The middle part of the game after the pieces have been developed, usually beginning around move ten and lasting until only a few pieces remain on the board.

Minor pieces. Knights and bishops.

Minority attack. An attack by several pawns against a larger group of pawns with the idea of weakening the larger group of pawns and opening up lines of attack.

Mobile pawns. Pawns that are not blocked and can move up the board quickly and easily.

Mobility. See piece activity.

Open file. A file not blocked by any pawns.

Open pawn structures. A position with no pawns blocking the free movement of the pieces.

Opening. The first ten or so moves of the game when most of the pieces are developed and the king is castled.

Opposite-colored bishops. A situation when each player has one bishop remaining on the board, with each bishop traveling on different colored squares where they cannot attack each other.

Opposite side castling. When one player castles on the kingside and the other player castles on the queenside.

Outpost. A strong square, usually in the opponent's territory, where a piece (usually a knight) can be safe, protected by a pawn, and cannot be attacked by an enemy pawn.

Outside passed pawn. The passed pawn furthest away from the other pawns where the main battle will take place.

Overworked piece. A piece that has to protect two pieces (or squares) at the same time that can often be taken advantage of tactically.

Passed pawn. A pawn that can move all the way up the board to its promotion square without being blocked or captured by an enemy pawn.

Passive. Holding back.

Pawn arrow. A line from the player's side of the board toward his opponent's side of the board of a pawn chain that is locked up on two or more files. Pawn arrows point in the direction where the player has more space and generally wants to attack.

Pawn chain. Pawns on files next to each other that are connected in a diagonal line, so they protect each other.

Perpetual check. The most common type of threefold repetition, when the player who is usually losing the game forces a position where he can check the other king back and forth forever.

Petroff's defense. A chess opening with a position arising after 1.e4 e5 2.nf3 nf6.

Piece activity. Having pieces on good squares where they can move freely.

Pin. When a long-range piece (queen, rook, or bishop) attacks an opponent's piece that is shielding another piece of greater value.

Process of elimination. A decision-making process where, to find the best move, you eliminate legal moves that you determine are clearly not the best move.

Promotion or promoting a pawn. When a pawn gets across the board to the eighth rank and turns into another piece, usually a queen.

Protected. When a piece is defended by another piece, usually making it a bad idea for the opponent to capture it.

Protected passed pawn. A passed pawn that is protected by a friendly pawn.

Queenside. The half of the board where the queens begin the game (the files a, b, c, and d).

Queenside majority. Having more pawns on the queenside than the opponent.

Ranks. The chess name for rows that run horizontally across a chess board.

Resign. A way to lose a chess game, typically signified by tipping over one's king into a hopeless position.

Rook lift. Moving a rook forward, usually to the third rank, and then in front of pawns on the second rank, where it can attack a file into the opponent's territory.

Ruy Lopez. A chess opening with a position arising after 1.e4 e5 2.nf3 nc6 3.bb5.

Sacrifice. Giving up material in order to gain some other type of advantage or checkmate.

Sicilian defense. A chess opening with a position arising after 1.e4 c5.

Skewer. A type of tactic like a pin, but where the more valuable piece is in front of the less valuable piece.

Space. An element of chess related to piece activity, referring to the number of squares controlled by each of the players.

Square of the pawn. An imaginary square the defending king must get into in order to win the race to the promotion square against an enemy pawn.

Stalemate. A type of draw where the player to move has no legal moves.

Stonewall Dutch defense. A chess opening with a position typically occurring after 1.d4 f5 2.c4 e6 3.g3 d5 4.bg2 c6.

Symmetrical pawn structure. A pawn structure where on each file that white has a pawn, black has a pawn opposite him.

Tactics. Immediate threats and attacks that make up the battles between pieces.

Threefold repetition. A type of draw that occurs when the same identical position repeats three different times.

Tempo. A single move, relating to time.

Threat. An aggressive move that attacks an opposing piece.

Tripled isolated pawns. Three pawns lined up vertically on the same file with no pawn on a file next to them to provide protection.

Tripled pawns. Three pawns lined up vertically on the same file with a pawn on a file next to them that can provide protection.

Uncastled king. A king that has not been castled and is usually in the middle of the board.

Variation. A logical sequence of moves different from the main line.

Zugzwang. A German word meaning "compulsion to move." Usually occurring in the endgame when there are fewer available moves, a player is "in zugzwang" when it is his move, and every possible move makes his position worse. It is when a player would prefer to pass his move to his opponent if he could.

7th rank. The rank on the opponent's side of the board where his pawns begin the game.

8th rank. The rank furthest away from the player, where his pawns promote.

Great Players

Mikhail Botvinnik (1911-1995) world champion (USSR) – 1948-1957, 1958-1960, 1961-1963

Jose Raul Capablanca (1888-1942) world champion (Cuba) – 1921-1927

Irving Chernev (1900-1981) noted chess author and master

Bobby Fischer (1943-2008) world champion (USA) – 1972-1975

Hans Kmoch (1894-1973) Austrian chess international master, author, and journalist

Alexander Kotov (1913-1981) Russian grandmaster and author

Bent Larsen (1935-2010) 6-time Danish champion and world champion contender

Emanuel Lasker (1868-1941) world champion (Germany) – 1894-1921

Edmar Mednis (1937-2002) U.S. chess grandmaster and author from Latvia

Paul Morphy (1837-1884) unofficially the first world champion from New Orleans

Lajos Portisch (born in 1937) 8-time Hungarian champion and world champion contender

Rudolf Spielmann (1883-1942) Austrian world champion contender and author

Siegbert tarrasch (1862-1934) German world champion contender and chess writer

Savielly Tartakower (1887-1956) 2-time Polish champion and chess journalist

Eugene Znosko-Borovsky (1884-1954) Russian chess master, teacher, and writer